COLOURFUL CACTI AND OTHER SUCCULENTS
OF THE DESERTS

By Edgar Lamb

THE ILLUSTRATED REFERENCE ON CACTI AND
OTHER SUCCULENTS
Volume 1 (1955, 1958, 1963, 1973)
Volume 2 (1959, 1974)
Volume 3 (1963) (with Brian Lamb)
Volume 4 (1966, 1974) (with Brian Lamb)
STAPELIADS IN CULTIVATION (1957)
THE FLOWERING OF YOUR CACTI (1955)
CACTI FROM SEED—THE EASY WAY (1959)

By Edgar and Brian Lamb

POCKET ENCYCLOPAEDIA OF CACTI IN COLOUR
INCLUDING OTHER SUCCULENTS (1969)

EDGAR and BRIAN LAMB

COLOURFUL CACTI
AND OTHER SUCCULENTS
OF THE DESERTS

with 140 Photographs reproduced in Full Colour

BLANDFORD PRESS LONDON

© 1974 Blandford Press Ltd
167 High Holborn, London WC1V 6PH

ISBN 0 7137 0673 2

Colour printed by Colour Reproductions Ltd., Billericay

Printed in Great Britain by Page Bros (Norwich) Ltd, Norwich

CONTENTS

LIST OF ILLUSTRATIONS

The maps included in this book are intended as a guide to the places referred to, such as the habitats which were studied, the desert regions and the National Parks and Monuments. For the sake of clarity only those mountains which are mentioned in the text are shown on the maps.

PREFACE

For very many years now, there has been a gap in the reading matter available to cacti enthusiasts, particularly those who have never seen these plants growing in nature. We feel that this book exactly meets this need, enabling amateur enthusiasts to get a better understanding of the varied environments in which these often viciously spined, but beautifully flowered, plants live. As we are in regular contact with amateurs all over the world, we know about the many misconceptions that many of them have regarding their plants—including some who have been growing them as a hobby for a number of years.

This book has been written in a totally different manner from any other on the subject of cacti; its purpose is to give readers a true insight into how these plants grow in the wild state. Much has been written over a period of very many years, stating that these plants require full sun treatment at all times, little or no water and so on. It is true to say that many mature specimens grow in full sun, but how many amateurs possess only mature specimens!

We have selected in a quite arbitrary fashion a number of habitats from Texas through to California in differing types of locations. Then we have included illustrations at each location, first showing a more general scene, followed usually by a series of close-ups of the individual species found in that area, along with a few paragraphs about the location and the plants. This is why you will not find plants illustrated in any botanical or alphabetical sequence. Full indexes are included for easy reference.

Many may never have the opportunity to see these plants in nature, and the illustrations alone will give a much better idea of their native home. To a certain degree other cacti from Central and South America, and many of the other succulent plants from Africa have evolved in much the same way, growing in locations with a very similar appearance and soil, and under almost identical climatic conditions. Although this book is devoted to the desert or semi-desert areas of the USA, many of the points raised and answered apply to other unrelated plants which come within this wonderful hobby. South African Aloes for example have evolved in shape and form in much the same way as the Yuccas of North America.

We have also included the phonetic method of pronouncing the Latin names, which are the reliable way of identifying plants, rather than by common names, which often refer to more than one species. The chapter on cultivation is not intended to be complete, although we have included a useful section on growing plants by artificial light. Those who require much more detailed information on cultivation, should refer to our previous book *The Pocket Encyclopedia of Cacti in Colour*. We are certain that our hints on photography will also be of assistance, as we operate an identification service to subscribers, which is often made very difficult, if not impossible, through poor quality photographs and colour transparencies. Sometimes everything is blurred, or wrongly exposed, and yet close-up photography is quite easy if tackled in the correct manner.

We have emphasised the need for conservation, and this does not just apply to that part of the world, but everywhere. Even today many plants are illegally collected in the 'wild' and transported to another continent, where perhaps only 10 per cent will survive. If this continues, a time will come when a large proportion of these plants will be extinct, and future generations will not have the pleasure of their gorgeous flowers.

In a book of this size it has not been possible to include illustrations of every species that could be found within the desert regions and National Parks of the USA, but we have included simplified descriptions of at least one species from each genus. We have also included certain other succulent and xerophytic trees which are also a prominent feature of the desert regions.

1. WHAT IS A CACTUS—OR SHOULD WE SAY A SUCCULENT?

Strictly speaking, it is incorrect to refer to all the species of plants illustrated and written about in this book as cacti, but it would be correct to say most of them are succulents. The word succulent is not really a good botanical term, but is generally used as a means of describing those plants which have water storage tissues, whether in the form of succulent stems, succulent leaves or a succulent root system.

These plants could also be termed xerophytes, in other words plants which have adapted themselves to live in dry areas where the rains can be few and far between, or where they have to make use of sea mists as with some of the coastal species. Not all xerophytic plants and trees have succulent tendencies, including a few of the plants mentioned in this book. Instead of developing areas of water storage tissues, such as swollen stems, leaves, etc., they have evolved in a different manner in order to withstand dry conditions, by reduction of the leaf surface area, only having stomata on the lower surfaces of the leaves or in sunken pits in the leaf surfaces. In some cases the leaves may roll up in times of drought and this is another means of reducing the transpiration rate. Almost all xerophytes, including the succulent plants, have a thickened leaf surface or epidermis, which in very many cases has a waxy finish to it; these are aids to the reduction of water loss from the plant.

In this book you will also see mentioned various shrubs which have xerophytic qualities in order to grow where they do. In many cases these xerophytic bushes shade some of the smaller cacti for their complete life-span, or, as with the well-known **Saguaro** Cactus (*Carnegiea gigantea*), for the juvenile period.

Cacti are a family of plants known as Cactaceae which, considering the age of this earth, are a very new group of plants, probably little more than 20,000 years old. True fossilised remains of cacti are not known, because of their recent evolution. Cacti are an exceedingly varied group of plants because of the way in which they have adapted themselves to suit the varying environments where they are to be found. This variability in adaptation is quite remarkable considering that it has happened in perhaps 20,000 years.

There are only a few cacti which are leaf succulents; they come from the West Indies and are outside the scope of this book. By far the majority of cacti are stem succulents, in other words they possess a very well-developed stem containing plenty of soft tissue which is capable of storing water. The surface of the stem usually has a waxy outer surface or epidermis, which tends to reduce water loss. With their woody core and rib structure, cacti are so constructed that they can literally swell when water is being absorbed, yet do not collapse in times of drought but only shrink to a limited degree. There are also some cacti such as *Peniocereus* and *Wilcoxia* which have swollen tuberous root formations.

Most cacti are clothed in spines to some degree. These can serve a number of purposes. The more densely covered a cactus is with spines, the better protected it is against the sun, as they cast a shadow over the stems in much the same way as the laths do in a lath-house. Moisture can also condense on the spines, so that if the spines point in a downward direction the drops so formed will drop off and be made use of by the roots, which with many cacti are very near the surface of the soil. Some cacti such as the Cholla-type Opuntias have easily detachable joints, barbed spines which can catch in the fur of animals and enable the joint to be transported for miles before being rubbed off, when it will grow into a new plant. Spines are actually very unusual leaves; they vary in size and shape considerably and appear from areole positions. This is explained clearly in the Glossary of Botanical Terms (p. 228) which includes some diagrammatic drawings.

You may be wondering by now how to know if a plant is a true cactus. The separation of one family of plants from another is basically through the flower structure. With many kinds of plants, it is possible to distinguish clearly between the sepals and the petals through the structure of the flower. With cacti this is not always possible, as they intergrade with one another. A cactus has the usual stamens and stigma, but it has an inferior ovary as the upper part of it is joined to the petals. The seeds which form later in the fruit or berry are in one cavity. The fruits are variable in size and so are the means by which they dispose of their seeds. Some fruits, because of their colour, their smell when ripe, and fleshy structure, are eaten by animals and birds, and these then scatter the seeds at a later date. Other fruits are dry or may have a basal pore so that when it ripens the seeds flow out of the bottom.

Other succulent plants described which do not belong to the family Cactaceae such as *Yucca*, *Agave*, *Dudleya* and *Graptopetalum* are all leaf succulents. In addition we have included such genera as *Dasylirion*, *Fouquieria* and *Nolina*, which are very xerophytic but do not possess as much succulent water storage tissue, as they are prominent members of the desert flora of the south-western part of the USA. These belong to other families, as listed in Chapter 11.

2. GEOGRAPHICAL DISTRIBUTION OF CACTI AND CLIMATIC VARIABILITY

True cacti are New World plants, with the exception of a few epiphytic *Rhipsalis* which have been found in Central and East Africa. Members of the cactus family are to be found as far south as the southern tip of South America and in a northerly direction in Canada, also in the tropical islands of the West Indies and the Caribbean, and the Galapagos Islands in the Pacific. They are found in coastal areas within a few feet of the sea, in jungle areas and in high mountainous areas such as the Andes Mountains where a few species have been found above 12,000 ft (4,000 m). With such a variation of habitats and climatic conditions, some species are only to be found in certain places, but there are species of cacti that have an equally wide range from hot, dry desert regions to the colder areas a few thousand feet up, where they have to endure snow and freezing conditions in winter.

This book is an introduction to cacti and some of those other succulent plants growing in the USA which not only consists of a very wide area, but also has considerable climatic differences. Cacti are to be found as far east as the Atlantic seaboard in Virginia, further north in New Jersey, where fairly severe winter conditions occur, and down into Florida where in the south the climate is termed as being of the savannah. Here the rainfall is quite high, sometimes in excess of 50 in. (127 cm) per year, and this occurs in the warmer months of the year. The winters are fairly dry, but far from cold, and some of the cacti to be found here are quite different from those found in more desert-like regions. Many of the species from a savannah climate are quick growing and very prone to frost damage, which occurs only on rare occasions in southern Florida. Cacti in their wild state are to be found in all except three of the states of the USA, including those bordering on Canada, such as North Dakota and Minnesota, where the climate is very severe in winter, with temperatures well below zero (Fahrenheit scale). Snow will lie for a few months each winter, but the summers will be hot: this is known as a boreal climate. Quite a variety of cacti, as well as some of those other succulent plants, are to be found in the western states, especially in California, on the Pacific seaboard (also called the California Chaparral region) in a subtropical climate. A subtropical climate means that in winter there is a reasonable amount of rain and a short cold dry period when frosts will occur, while in the summer a smaller amount of rain is likely and temperatures up to 90°F (27°C) can be attained at times. The main growing period for plants here is in the spring.

The regions of the USA which possess most cacti and some of these other interesting succulent plants are southern Texas, southern New Mexico, Arizona and southern California. In these states there are vast areas of desert and semi-desert regions which abound in cacti. Some of these desert areas are in fact extensions of much larger deserts in Mexico, such as the Chihuahua desert which extends northwards to the south-western and western parts of Texas, north of the Rio Grande, which is in fact the border between Mexico and the USA. This includes the area to the west of the Pecos River often termed the Trans-Pecos region. This area in particular the Big Bend is a very rewarding one in which to study the desert flora, as you can see by the series of illustrations (Pl. 1 to 54), which were filmed there. Temperatures in summer are very high with little or no water, while in winter a few inches of rain should fall, and on occasions snow may lie for a few hours or days. A little rain may fall in late summer too. The average elevation in this Trans-Pecos region is between 3,000 ft and 4,000 ft (900 m and 1,200 m). So the plants and trees here have to be capable of enduring temperatures not only well in excess of 100°F (38°C) but also down to below freezing. Except at higher altitudes these below-freezing conditions will only be for a short period, and generally when the plants are dry. Plants on the whole will endure many degrees of frost if they are dry. There are in fact quite a range of 'hardy cacti', that is species which will stand temperatures down to zero on the Fahrenheit scale ($-18°C$) if they are dry. Some of these plants are to be found at much higher altitudes, perhaps growing either in Prairie grassland or in Juniper-Pinyon woodland, occasionally in excess of 10,000 ft (3,050 m). In order to cope with these conditions in winter, the plants shrivel. This has two effects: first it concentrates the sap of the plant, so that it lowers the point at which the sap can freeze, like the anti-freeze in a car; secondly, by shrivelling, each cell of the plant is only partly full of sap, so that if the temperature is low enough for it to freeze there is sufficient space for expansion, bearing in mind that a liquid such as water for example does expand on freezing. The plant would suffer if the cell walls were to rupture, causing the general rotting and disintegration of the plant. If plants endure many degrees of frost when they are turgid or full of sap, damage is almost certain to occur.

The Chihuahua and Sonora deserts extend into southern New Mexico and Arizona, although when in Arizona and California you also hear of the Arizona and Colorado deserts. These are in fact no more than the northernmost extensions of the Sonora desert of Mexico, at altitudes between 1,500 ft and 4,000 ft (450 m and 1,200 m). The average annual rainfall is 11 in. (28 cm) at the lower altitudes and up to double this amount at around 8,000 ft (2,450 m). The rains come in winter (November to March) and again during late summer (July to September). The Mojave desert is also at about the same elevation in San Bernardino county and eastwards into south-western Utah. The Mojave desert links up with the Navajo deserts at altitudes of around 5,000 ft (1,525 m) in the southern parts of Utah and Colorado, where again there are very hot dry summers but it is quite cold in winter, and certain 'hardy cacti' again occur.

3. HOW DO I NAME MY CACTI?

The problem of classifying or naming cacti is a big one. There are really three schools of thought, of which the basic one is the conservative school where there are a minimum number of generic names. The generic name or genus is a subdivision of a family of plants such as the Family Cactaceae which includes one or more species (different plants) that seem to have certain common characteristics and ancestry. The first name of the species is the generic name showing to which genus it belongs. Then there is the school of thought which splits the cactus family—and other families of plants for that matter—into as many genera and species as possible. Unfortunately, the tendency in recent years has been to do this, but we prefer to consider ourselves as belonging midway between these two factions, which basically means following the classification system introduced by Britton & Rose in their four volumes produced between 1919 and 1923. It is impossible to follow them exactly; for one thing some really new discoveries have been made since then, which have resulted in their being described as new species, and in some cases new genera have come into being as these new species did not fall into any previously known genera.

Another problem that occurs from time to time is the changing of the name of a cactus. You may wonder how this can occur.

When a new plant is to be described, the person who is naming it should try to make certain that he is not describing a plant which has already been named. If his previous research work has not been thorough enough it may already be named, and, generally speaking, by the rules of botanical nomenclature, the first published name is the valid one. However, in years gone by some plants were originally described in little-known publications, so it was possible a few years later for a plant to be described without knowing that this had already been done. This is one of the ways in which plant synonyms can occur. In Chapter 12, where the descriptions of the species are given, the better-known synonyms only will be mentioned.

Some cacti have common names such as Barrel Cactus, which refers to the genus *Ferocactus*, and Hedgehog Cactus for *Echinocereus*. Not all plants have accepted common names; and some common names may refer to more than one kind of plant. So use these names if you wish, but do not rely on them entirely as they can be misleading. It is far better to get to know the correct botanical names. They are not difficult to learn and pronounce.

Returning to the identification problem, considerable variability can occur within a species, and this is where the hair-splitting school of thought in some groups of plant life have had a field day, publishing still further species and varieties. In some cases this has been done without ever studying them in the 'wild state', but from receiving a few plants from the 'wild', or even raising them from seeds they have been sent. Many of these are only variations of older, well-known species.

Some species are far more stable than others, so that the variation from plant to plant is minimal not only in one colony, but also in colonies which may be separated by sometimes a few hundred miles. Some plants can have a very wide range from near desert to almost alpine conditions, and variations within a species are to be expected, as they have evolved in each locality according to the conditions prevailing. If, then, you select two plants, one from the desert location and the other from a much higher altitude, you could well think that you had, if not two distinct species, at least two good varieties. However, by looking at other specimens from intermediate locations you could note the gradual gradation from one to the other. Usually these variations are mainly confined to the vegetative structure of the plant.

Where a species is far less stable you can find plants in one location where the differences are very noticeable. For example, look at Plates 97, 98 and 99 of *Echinocereus fendleri* var. *boyce-thompsonii*. In this one location it was possible to find specimens varying from white spined to brown or even black. The spine lengths were fairly constant, whereas at a location for *Echinocereus fendleri* var. *bonkerae* the spines varied not only in colour but also in length and formation. The flowers however were fairly constant between plants of differing spine colour and length. Flower colour variations do occur, but they are not worthy of different names, and occasionally a plant has been found with yellow flowers on one side and pink on the other, all coming from one stem. This has been noted with *Echinocereus dasyacanthus* and other types of plants too. The shape of the petals can vary between plants of the same species, and double flowers occur with approximately twice as many petals.

It must also be borne in mind that not all plants take on their mature form straight away, so that some plants in their juvenile state can have a very different appearance to older specimens. For example, some of the Barrel Cacti in their first few years (up to around 8–10 years usually) have a tubercled appearance, usually a number of conical protuberances upon which the spines are borne. In old flowering plants these tubercles merge together to form a prominent rib structure, whereas previously the tubercles were in a series of rows only. Also with some species the spination can be different on young plants, perhaps no prominent central spines, or far fewer spines than on adult specimens.

The examples already mentioned are those where the young plant is incapable of flowering, the extra spines or centrals being a maturity sign, as flowers do not occur until they appear. What is more of a problem is where you have a juvenile form, perhaps without centrals, which is also capable of flowering. We are including an illustration of *Coryphantha echinus* (Plate 16) showing two distinct forms, one with strong central spines so well developed that the flower cannot open properly. The other form has radial spines only, and as there are no centrals the flower has been able to open out wide. At one time the juvenile form without any central spines was named as a separate species, or *Coryphantha pectinata*, this latter name referring to the comb-like spine formation. Today it is realised that these two plants are really one and the same plant, and the only question now is the cause of the difference. These plants were found growing side by side under identical conditions. We have in our Exotic Collection observed how some plants have reverted in growth to radial spine development only, and then in a year or so started to produce them again. Some authorities believe that a plant in poor growing conditions may do this, but we feel that there is still more to be found out on this subject.

It would be possible to go further into this question, but in an introductory book of this sort it is not called for and would only complicate matters still further. It is sufficient to explain that you will come across many variations, and we have given you a clue to some of the possible reasons. So you must not necessarily leap to the conclusion that you have found a new species if you cannot identify it, but we hope that this will not dampen your enthusiasm.

A simplified means of separating most of the genera will precede the descriptions of species section (Chapter 12), which, with the illustrations, should help you to identify many of the plants that are to be found in the desert regions from the Gulf of Mexico in Texas through to New Mexico, Arizona and southern California.

4. FREAK FORMS AND CRISTATES

From time to time in the wild and in cultivation something unusual occurs to cause an abnormality. Occasionally these abnormalities have been caused by damage; for instance a globular cactus which is normally solitary may have its top knocked off and from it may grow a number of new heads. This is not a freak form which can be perpetuated by vegetative reproduction. Sometimes curiously-shaped flowers may also be produced in this way.

In some cases abnormal growths appear which are caused by the normal genetic make-up of the plant being disturbed. This type of abnormality can appear in many ways, but where it distorts the shape of the plant it is possible to perpetuate such a freak plant by dividing the plant or removing these unusually shaped parts of the plant, rooting them and growing them on. As a general rule when this occurs the plant tends to try to revert to normal, so if you wish to perpetuate a freak form you remove the normal parts of the plant, then the energy or all the growth available goes into the freak sections.

These freak forms can take on many shapes, but usually they are referred to as monstrous or cristate forms. The monstrous forms are irregular shapes as a general rule, whereas a cristate form is where it appears as though a number of stems have flattened and joined together. Another term applied to this is fasciated. Some of these cristate or fasciated forms can be quite attractive, sometimes forming into undulating mounds. Instead of having a single growing point on a stem, you have a strip of cells or growing points, to put it simply, causing these cristates. The real cause of freaks such as this is unknown. Over the years various people have claimed that by damaging or burning the growing tip it was definitely possible to make the cristate forms occur. So far this has always been disproved.

Plate 71 of *Carnegiea gigantea* shows a very fine cristate head on this giant cactus. A plant such as that is a rarity, whereas *O. fulgida* var. *mamillata* fa. *cristata* is quite common in collections. A colony of this cristate form is to be found east of Florence, Arizona. If seed is sown from fruits off a cristate plant, the chances of cristate seedlings appearing are only slightly greater than from seed off a normal plant.

5. APPRECIATING THE DESERT AND CONSERVATION

There are a number of definitions for the word 'desert' such as a 'waste solitude' or a 'barren tract of land', and to many people this would mean the last possible place on earth which they would wish to visit. However, not all deserts are entirely barren. Plants and the many forms of animal life live in them, though they have had to adapt to the conditions in order to survive, as the heat by day during the hottest part of the year may reach 120°F (49°C) in the shade. Full sun temperatures would be a great deal higher. The amount of rain which will fall annually in these desert or near desert regions may be no more than 5 in.–10 in. (12·5 cm–25 cm), and some years little or no rain may fall at all.

It is not surprising therefore that much of the plant life written about in this book tends to be somewhat slow growing. This does not apply only to the large growing cacti such as the Saguaro Cactus (*Carnegiea gigantea*) where average specimens may be 200 or 300 years old, but also to some of the smaller cacti such as the Living Rock Cactus (*Ariocarpus fissuratus*) from Texas where even quite a small specimen perhaps 2 in. (5 cm) may be 50 years old. To a certain extent anyone who is familiar with these plants is making an educated guess, when it comes to age. We have been able to study plants in our Exotic Collection under almost ideal growing conditions for a period of some forty years, including specimens of species illustrated in this book. Despite the fact that our growing season is shorter, the growth rates of the two species mentioned above seem to compare quite favourably with specimens in the wild state. We are sure that *Ferocacti*, for example, grow a little quicker in Arizona than under glass; even so specimens a few feet high will be a great age. The present laws in the USA make it illegal for anyone to collect cacti and many other kinds of native plant life without permission from the State, and the landowner where they are on private land. Even when plants are on private land, State permission is still required. Despite this many areas of south-western USA have been stripped of cacti and a few species may well become extinct within a very few years. One reason for this is over-collecting. We have ourselves been very sad to see bins of cacti lying in full sun and destined to die long before reaching an amateur collector who would greatly treasure them. Modern developments of new cities and new roads also encroach on the wild areas where cacti flourish. However, many species do stand a very good chance of existing for a long time yet, with the laws as they stand today.

It would be a tragedy if they were to disappear entirely, because in spring even with little or no rain many cacti will still flower. You can see almost every colour and shade in cacti flowers except blue; some have a waxy texture, and can vary from a fraction of an inch to perhaps 12 in. (30 cm) in diameter. Few members of the plant kingdom can be so varied and beautiful and yet they come from such arid parts of the world. As you can see by the illustration, these desert areas are not necessarily miles of fine sand but often contain most unusual rock formations, in places almost likened to a lunar landscape.

Those visiting these parts for the first time, particularly in April or very early May, will almost certainly agree that these regions have a unique beauty. If there has been sufficient rain some weeks prior to your visit, you will see the added beauty of annual desert plant dwellers blooming within a short period, before the excessive summer heat arrives. Their seeds may have remained dormant for a few years awaiting the right conditions for germination.

Cacti and the other desert dwellers can be raised from seed, which can be bought from various firms throughout the world. You may also collect your own seed in the wild, but do first make certain that you are not violating private property or breaking any laws. Usually if you ask permission, you will get it, and invariably by doing so you will be put in touch with similar-minded people who share your interest in these wonderful plants, for there are numerous societies and organisations in these states. You may also be told of other places of interest nearby where you can see a fine display, perhaps in one of the botanical gardens, national parks or national monuments of which there are some devoted to the preservation of plant-life. The Desert Botanical Garden of Arizona near Phoenix, Arizona, is one such place. This was started on land where cacti grew naturally, and has been added to, so that it includes desert plant-life not only from the Americas, but also from other parts such as Africa. Another is the Sonora Desert Museum in Tucson, Arizona.

We have mentioned that modern developments can mean the end of many areas where cacti and other plants grow. Fortunately today there is more co-operation between departments so that where these have to take place, botanical gardens and sometimes commercial firms are given permission to clear the area. If this is done there is a reasonable chance that many of these plants will survive in captivity, and seed be produced and distributed to other botanical gardens as well as to amateur enthusiasts. There are amateur collectors in many parts of the world, who are playing an equal part in conservation. We have been doing this for many years in our Exotic Collection, and have in some cases been able to supply seed or plants to a botanical garden of a species which cannot be found in the country of origin, where that botanical garden was situated.

If it had not been for the assistance of certain landowners in Texas (who wish to remain anonymous) and the Arizona State authorities, it would not have been possible for us to collect a few specimens of each species for the purposes of filming, so this book would have been more difficult to produce. Their co-operation has also resulted in our being able to obtain seed by careful hand pollination, to avoid hybridisation, which we can distribute either in seed form or as plants.

There is a need today for the preservation of plants and animals in habitat, not only in semi-restricted areas, but also in private or public botanical gardens and zoos.

Plants do not disappear through over-collecting only, but also through wilful damage or vandalism. People sometimes carve their names on plants and this is one way which can cause a plant to die. If this is done at a period of year when rains will follow, those cuts will not callous or heal, so that infection gets into the plant and within a few weeks or months a cactus that has taken perhaps a hundred years to grow will be just a rotting heap. The careless lighting of fires can be an even greater menace, whether in forest land or in the desert regions anywhere in the wild, unless the proper precautions are taken. So let us all make certain that all forms of wild life, no matter whether they are in the plant or animal kingdoms, continue to thrive in the wild if possible or else in captivity.

6. WHAT CAN WE LEARN FROM CONDITIONS IN NATURE THAT HELPS IN CULTIVATION?

You will have already gathered that cacti are to be found usually in places where rainfall is rather limited, and that they have adapted to this in a number of ways. One is the shallow but spreading root system which the majority of plants have, which has enabled them to make use not only of the smallest fall of rain, but even of dew, or the condensation of atmospheric moisture on the spines, which just moistens the soil. Only a few cacti are really deep rooted, so in cultivation if they are to be grown in containers a shallow but wide pot is beneficial, to allow more normal root growth. Free root-run conditions are even better, as perfectly normal root growth is then possible, either in a greenhouse or in the open air, if you live in the right climate.

Under glass the growth rate which will be achieved can be quite fantastic, even if a number of specimens are planted quite close together, as in cultivation no matter whether indoors or out, plants are regularly watered instead of waiting for the next shower of rain. Plants in the wild, particularly where the annual amount of rain is small, will tend to be spaced out to a certain degree. If specimens were too close together growth would be much slower, as there would be less water for each plant. It is quite noticeable with some species, where they are the dominant form of vegetation in a particular area, how they tend to be almost evenly planted. This can sometimes be observed with a hillside of Saguaros (*Carnegiea gigantea*) in Arizona or with *Dasylirion leiophyllum* (Sotol) in Texas.

Some of the taller cacti such as *Carnegiea gigantea* will grow very well as pot plants up to perhaps ten years or so of age, then the growth-rate gets slower. This is due to their inability to develop their spreading root system, rather like the spokes of a wheel, in the confines of a pot. *Carnegiea gigantea* is in fact a species which will only develop properly as it gets larger under free root-run conditions. It is possible to grow it in a pot, but if growth nearly ceases do not be surprised. In effect you are then treating it in much the same way as the Japanese miniaturise numerous tall growing trees, which they then call Bonsai trees.

It will already be obvious that the cacti and most of the other plants dealt with in this book are not likely to need an excess of water in cultivation. Although it is true that cacti and these other plants have to endure long dry periods in nature, one cannot exactly term these growing conditions as ideal. In cultivation, no matter where they are grown, these plants will appreciate some extra water at more regular intervals during the growing months of the year, usually from spring to early autumn, but that does not mean that the soil should never be allowed to dry out. The quickest way to kill most of these plants is for the soil to stay too wet for too long a period, particularly around the neck of the plant. In cultivation you will often see coarse sand or grit as a layer on the surface of the soil, and this is the reason.

During the late autumn watering should be reduced and during the winter stopped entirely particularly if you live in a temperate climate. If you are living in a part of the world where you can grow most of your plants in the open, because frosts are rare or never occur, the winter resting period will be much shorter. In this case watering may be reduced, but it is usually better not to stop it completely.

A great deal has been written in various books and magazines on the soil requirements for cacti, some stating how exact one must be with the soil mixture as regards its pH (degree of alkalinity or acidity). Only a very few species are ultra fussy over soil requirements. By far the majority of species will grow quite well in a soil made up of about equal parts of a sharp sand and any good humus such as leafmould or peat. A well broken-down leafmould is the best from oak or beech trees, but in many parts of the world this will not be available, so use the nearest substitute to it. Such items as gypsum, superphosphate and bone meal can be added, but in small quantities. Certain proprietary compounds are also available for use with various types of plant life, which contain the various trace elements, to complete your soil mixture. Even without the addition of these extras the majority of plants will grow well in a sand and leafmould mixture. So you can completely ignore anything you may read which says that you have to be very exact over the soil mixtures for different types of species, or even different species within one genus.

In the wild these desert forms of plant life, even where the same species is concerned, can sometimes be found growing not only in climatically different habitat, but also in different types of soil. Even where a species is growing in an area of almost identical climatic conditions, the soil can differ. An example of this is mentioned in connection with *Neolloydia conoidea* (p. 44). Seeds of these plants can be scattered far and wide, falling on areas far apart, but being the tough plants that they are, they can survive wherever they fall, provided it is not where moisture can lie for too long a period.

When it comes to cultivating these plants, light is another important question to be studied. You will see in some of the habitat pictures that only a minority of species grow in the open in full sun conditions, and even those larger species would not be in existence if they had not had some shade in their early years. It is almost a physical impossibility for a seed to germinate in a full sun position and survive (with the odd exception). So when it comes to growing any of these plants from seed, the first thing to remember is the need of shade prior to germination, and also for some months afterwards. The degree of density of shading required as plants grow older can be reduced, until one comes to the question—do they need a little shade at all times? The majority of the small plants are usually found in partial shade beneath other trees and shrubs, and some species are capable of pulling themselves down level with the surrounding soil in the heat of summer. At this time of year it would be dry, and any loose sand or soil or even dry leaves in a few cases would drift over them, and give added protection from the sun.

Seedlings that have been raised under a glass which tends to cut off some of the ultra-violet rays will need acclimatising before being placed outside in full sun conditions, otherwise they are likely to burn. Today plastic, polythene, etc. in various forms is more commonly used for greenhouse structures, and this allows a far greater amount of the ultra-violet rays through, so that the bodies are tougher and less likely to be burned when put in the open. When it comes to the temperature that these plants will stand without damage, it is safe to say that plants outside will stand far higher temperatures in a full sun position than those under glass in a similar position. Most species will stand similar temperatures even in excess of 120°F (49°C) in a lightly shaded position and higher. Another important point to bear in mind is the angle at which the sun's rays meet a plant. For example, an erect specimen of *Carnegiea gigantea* (Saguaro) has to endure in nature at times sun temperatures of about 150°F (70°C) and does so without any damage being incurred. If through wind a specimen has been pushed over, burning occurs very quickly on the side of it now facing the sun. This same sort of thing can occur occasionally with plants on a shelf near the glass in a greenhouse, because of the angle at which the sun's rays strike the plant.

You might imagine from the very dry atmospheric conditions normally existing during much of the year in these desert or near-desert regions that in cultivation these must be repeated. They need only be repeated during the winter resting period, when temperatures are lower. Most kinds of plant life will survive quite low temperatures during the winter, if kept fairly dry at the root and accompanied by a low degree of atmospheric humidity. In a climate such as that of Britain, it is often safer to keep certain plants at a slightly higher temperature in the winter because of the high humidity. Those same plants in habitat endure quite low temperatures in winter, but accompanied by a low degree of humidity. A few plants even though their roots are dry can still rot, if the humidity is too high in winter.

However, in the warmer months of the year, when the plants are growing, many of them will appreciate a much higher degree of humidity than they would normally get in nature at that same period of the year. Obviously this reference to humidity is only applicable in cultivation where the plants are being grown in a greenhouse or some similar enclosed environment. Those who have a suitable climate and grow these plants in the open air cannot control the degree of atmospheric humidity to any real degree. However, early morning watering of the plants will help for a short time; in hot climates watering in the evening is probably the best time of day.

In nature, assuming that they have not been eaten by rodents and other such creatures, seeds will remain dormant for many years, until the right conditions come along. This means usually that the seed coat (testa) is quite thick, hard and impervious to water. If it were not, the contents of the seed would dry up, so that it would quickly become useless or non-viable. Because of this these seeds need plenty of water to make them germinate, sufficient to penetrate through the hard coat. In nature seeds lying on a sandy soil will blow around, and tend to be roughened by so doing; this can help break down the seed coat and allow water to enter more easily. In other cases where winter frosts occur, seeds can be cracked open by freezing, so this is another method by which nature helps. If seeds have been eaten by birds and animals, which were after the flesh of the fruit rather than the actual seeds, the acidity in the digestive juices can also tend to break down the seed coat and allow water to enter more easily. So in cultivation these natural methods can be imitated, for instance by scratching any very hard seeds or using a piece of sand-paper or emery-cloth. The freezing method is best carried out after the seeds have been sown in a seed-pan and soaked with water, and then placing the entire container in the freezer for a few hours. After this, remove it and treat in the normal way for seed-raising. The seed coat can also be broken down by soaking seeds in dilute acid, such as vinegar (dilute acetic acid) for a few hours, remove, wash clean, dry and then sow the seeds in the normal way. These are just three small hints from nature which can help with hard-coated seeds, which would otherwise be difficult to germinate.

So, as you will have gathered from this chapter, we can learn a lot from nature, which can help with the cultivation of these plants, but it does not mean that we have to try to copy exactly the same conditions.

7. FURTHER CULTURAL ADVICE FOR DIFFERING PARTS OF THE WORLD

The previous chapter has supplied a lot of useful information on the needs of cacti and most of the other desert plants for successful culture, no matter where you live. The basic requirements are the same as regards soil, water, light etc. no matter whether the plants are being grown in a tropical or near tropical garden, in a greenhouse or conservatory, or as house plants on sunny window sills. We are including here some additional information and suggestions under separate sub-headings.

TROPICAL CLIMATES

Even these can vary a lot, particularly when it comes to rainfall, either as regards the amount of rain annually, or the time of year when it comes. If you live in a climate where the annual rainfall does not average much more than 10 in. (25 cm) per year, this can be regarded as comparable with the conditions these plants experience in West Texas, Arizona and other areas in the USA. It will mean that the soil mixture can contain a much greater percentage of humus or another moisture-retaining substance such as Perlite or Loamalite. The tendency with these latter substances, which are artificially made, is that they are very light-weight and can blow away easily, and so are not suited for use in the open air. A good leafmould or peat is better in this case, and the former substance will contain nourishment value too. So in the rather dry climate we have been talking about your soil mixture could either consist of 2 parts by volume of humus (i.e. leafmould, peat, etc.) and 1 part gritty sand, *or* 1 part humus, 1 part good sifted loam, and 1 part gritty sand. However, if you live in a tropical climate where the rainfall annually can be very high 30 in. (75 cm) upwards, the atmospheric humidity will be correspondingly high, at any rate during the wet season. When this occurs, the soil can easily absorb enough water from the atmosphere for many plants to grow, as they do in habitat under similar conditions. So in order to combat this partially, a much better drained soil is required, as it will tend to reduce the amount of moisture which the soil mixture can absorb from the atmosphere. A soil mixture of either 2 parts by volume of gritty sand to 1 of humus or 4 parts gritty sand to 1 of humus and 1 of sifted loam is likely to be suitable. It is only possible to give a guide as regards the soil mixture for your area, but you will soon find how you can best adapt these ideas to your own particular conditions.

It has also been mentioned that some of these desert plants appreciate some shade, either in their juvenile stage or at all times, as with some of the smaller-growing kinds. This can be done by a number of methods, such as growing those species that need it beneath other trees and bushes, by means of a lath-house or even nylon netting shields. A lath-house consists of a light-weight wooden structure with a flat top, on which are nailed wooden laths ½ in. or 1 in. (1·25–2·5 cm) wide. As the sun goes across the sky these laths cast a moving shadow on the plants. The degree of density of shade can be varied according to how wide the spaces are between the parallel rows of laths. Alternatively, you can today obtain various types of nylon netting, or netting made of some similar synthetic material, where the size of the mesh varies. A closely woven netting will give most shade. These materials are also easily erected temporarily or permanently on poles and wires to give the required shade. We would not like to hazard a guess at the life of these materials in the full sun, but they are certainly easier and quicker to erect than the old lath-house.

Finally, we should mention that plants growing in the open air in some parts of the world do at times have to contend with rather larger hailstones than we are generally used to in temperate climates. One inexpensive way of avoiding damage to your plants is to use clear polythene sheeting (preferably a heavy-grade) in between two layers of wire mesh. The mesh size of the wire netting needs to be small enough, so that any hailstones will bounce off. Polythene sheeting can be obtained in large rolls and is relatively inexpensive, but it does become brittle with continued exposure to sunshine, which contains ultra-violet rays. It will be necessary to renew the lengths of polythene sheeting between the wire netting annually. It is quite an easy matter to make sloping roofs for cacti beds and rockeries by this method, and this will not only protect the plants from hail, but also from too much rain. Corrugated fibre-glass can also be used, but the initial cost is far higher.

All the plants referred to in this book will winter satisfactorily under these polythene shelters or in the open, and many of them stand a few degrees of frost if dry, some more than others. If you intend growing cacti and other succulent plants from more equatorial regions, where frosts are unheard of, you will then need a greenhouse or glassed-in verandah, perhaps with some form of heating.

TEMPERATE CLIMATES I : OUTSIDE FOR FROST-RESISTANT SPECIES

It is possible to grow a limited number of species of cacti and the other succulent plants in raised rockeries without overhead protection from frosts and snow. The species which can be grown in this manner are those which hail from the colder regions of the USA, and just a few extend into Canada. The greatest danger with any of these plants is excess moisture in the resting period. In order that these plants can endure freezing they must be able to shrivel a little, so in some places where there is high winter rainfall, a little overhead protection is preferred. Many species of *Opuntia*, *Neobesseya*, one *Pediocactus*, the odd *Echinocereus*, at least two species of *Agave* and most *Lewisias*, all native to the USA, will safely endure temperatures down to o°F (−18°C) and some even lower than that, and often covered with snow for many weeks at a time. So to succeed with such a venture a raised rockery of some sort or other is needed, with plenty of old bricks, stones and gravel at its base, and then topped off with a very porous soil mixture, made up of 4 parts by volume of grit and sand to 1 of humus. Select a sunny situation, and if it can be backed by a garden wall or the wall of a house or garage, all the better.

TEMPERATE CLIMATES 2: INSIDE A GREENHOUSE OR CONSERVATORY

This method of culture in a temperate climate is really the easiest way of growing desert plants, as it is a relatively simple matter to control the degree of moisture, light intensity and temperature required. Most people consider that a greenhouse, no matter what type it is, should be running in a north–south direction, but this is not absolutely essential. When it comes to a conservatory, a southerly aspect is ideal, but quite a good range of plants can be grown successfully in east or west facing ones. Where this occurs it may mean that a few species which need maximum light should be avoided.

We favour when possible growing plants under free root-run conditions, but this does not mean that you cannot take your plants to flower shows, etc. Clay pots have been in use for a very long time, but today plastic pots are more popular. For the majority of species plastic pots are quite suitable, but they hold the water longer, as they are not porous like clay pots. We feel that there are a few of the smaller, very slow growing species, such as *Ariocarpus fissuratus* (Plates 37 and 38) or *Epithelantha* (Plates 39–41) which are best grown in clay pots. It is certainly not impossible to grow them in plastic pots, but much greater care is required to avoid over-watering, otherwise rotting can take place. So, with plastic pots, in addition to ensuring that there is good drainage around the hole at the bottom, a slightly greater percentage of gritty sand may be needed. We favour for pot culture a mixture of about equal parts by volume of humus to gritty sand, but this can be varied according to the plants being grown. A mixture such as this will not set hard, whereas one containing sifted loam will tend to set much harder after a few years. The same thing can also occur if the sand used is not well washed. We prefer to use a pot sufficiently large to allow the plant to remain in it, undisturbed for perhaps as much as 4 or 5 years, rather than potting on every year or so as is often recommended. Remember, also, as stated in the previous chapter, that a wide pot is preferable for many of the species rather than a deep one.

As a general rule tap-water can be used for watering, unless it is known to contain any impurity or added substance which could prove harmful to plants, even though it may be safe for humans. Local knowledge about this is usually available. Always water your plants overhead with a spray as this helps to keep plants cleaner and healthier; this advice applies generally, no matter where the plants are grown. The best time to water is always the late afternoon or early morning rather than in the heat of the day.

The temperature in a greenhouse or conservatory can become very high and be more harmful to some plants than the same temperature in the open air. Harm is more likely to occur through unshaded glass, as some plants can burn, resulting in scars of a permanent nature. A temperature under glass or plastic around 100°F–110°F (38°C–44°C) is quite safe and can be even higher if shaded. One way to reduce the temperature is by ventilation, but we believe it is not a good thing if it means turning the greenhouse into a wind tunnel. If you do this, the plants will in some cases be transpiring moisture through their stomata faster than the roots can take it in, and you will need to live with the watering-can or hose-pipe in your hand!

There are two methods of shading a greenhouse: by spraying a substance on the outside, or using blinds which can be lowered during the hottest part of the day. The spraying method can be carried out either by using a proprietary brand of greenhouse shading or by mixing some powdered whitening with water, then adding some unused car oil to it. This latter addition helps it stick to the glass better, so that the first shower of rain does not wash it off. Blinds can come in many forms, and if you prefer this cleaner but more expensive method, you should consult a local horticultural firm.

Now we come to the question of heating the greenhouse, which is necessary unless you live in an area where the frosts are slight in winter. We reckon that most of the plants written about in this book would be quite happy in winter, even if the temperature fell to 40°F (5°C) and in some cases lower. This is always assuming that the plants are dry and in the dormant stage. There are many methods of heating the greenhouse or conservatory, from an extension of the household central heating system, to gas convection heaters, electric fan heaters or an ordinary paraffin heater. All these methods are quite suitable; the last method is usually reckoned as the cheapest. Those who are living in an area where sub-zero (Fahrenheit) temperatures occur frequently in winter may find that double-glazing is essential in order to maintain a reasonable temperature. Alternatively, in areas such as these, it is often cheaper to transfer your plants to the house or cellar for the worst months of the year. If they are in a cellar, artificial lighting will be required even though they are in the dormant state.

TEMPERATE CLIMATES 3: AS HOUSE PLANTS

The last two illustrations in this book depict two attractively arranged displays of plants, all of which come from the USA, both growing in containers which do not possess drainage holes, one being pottery and the other plastic.

If you wish to grow plants successfully indoors it is necessary for most species to receive the maximum light possible, either natural light or using artificial lighting, as mentioned a little earlier. If the plants are grown on window-sills, the light will usually come from one direction, and it is preferable to turn them round every few days so that the light is distributed evenly. If this is not done, most plants will tend to grow towards the light, and if in poor light, they will become etiolated. This means thin, weak and very unhealthy, unnatural growth.

The easiest way to grow plants indoors is by using normal pots with drainage holes, as mentioned in the section on greenhouses. If you use containers without drainage holes, great care must be taken at all times with watering, so that the soil does not become saturated to such an extent that it is like a bog. If this happens the plants will soon rot and die. If you wish to plant up containers, similar to those illustrated, put in plenty of broken pot and small stones at the bottom, before filling up with the soil mixture, which should be of a rather more sandy nature. Also it is better if you use plants of similar growth-rate in the same container, otherwise the quicker-growing plants will take over and smother the other slower-growing plants. The only trouble is that with plants of even growth-rate it is not always possible to obtain the best decorative effect, so we can suggest an alternative method. Provided the containers are large enough, you could have the plants growing in individual pots, and then disguise these by filling up between the pots and over them with small stones, and the occasional piece of rock. This will mean that the plants can be watered individually according to their needs, and when the quicker-growing plants get too large, remove them and substitute others. It will also be easier to control the amount of water given and avoid any rotting troubles caused by over-watering.

No matter which type of method you are using for these bowls, you may wonder when to water. One means of helping to discern the dryness of the soil mixture is to put a $\frac{1}{2}$-in. or 1-in. (1·25 cm) or 2·5 cm) diameter piece of tubing in a vertical position somewhere in the bowl, disguised by a small piece of rock. This tube, which can be of metal or glass, will contain no soil, but will go down on to the bottom drainage layer at the base of the bowl. By this means you can see if the drainage material is reasonably dry; it certainly must not be swimming in water. If you find there is water remaining there as a layer for more than a few hours after watering, the excess can be removed by a number of methods. The bowl can be tipped on one side carefully or, better still, soak the water up by capillary action using a length of cotton wool. To do this one end of the cotton wool should be at the base of the bowl on the inside and the other outside the bowl, hanging below the base of the bowl. It will only take an hour or so for the excess water to disappear.

We have suggested that watering should be done as an overhead spray, but this may not always be possible with plants indoors. But it is worth the effort at least three or four times each year to remove the plants no matter whether they are in pots or bowls to the garden or balcony where they can be given a good overhead spray.

As central heating is very widespread these days, the tendency will be for house plants to be kept in rather warmer winter quarters than they really need. If this cannot be avoided, it will mean that some watering will be necessary during the winter months, but it should only be enough to keep the plants turgid rather than growing too much. If your plants can be wintered in a room where the temperature falls to 45°F (7°C) or a little lower, they can then remain dormant for much of the winter, as with plants grown in greenhouses or conservatories.

TEMPERATE CLIMATES 4: INDOORS USING ARTIFICIAL LIGHTING

Today it is possible to obtain tubular Grolux lights under which plants can be grown, even to the extent of the plants never having natural daylight at any time in their life. These lights can be obtained as double fluorescent tubes complete with reflectors, 48 in. (1·2 m) long. These should be suspended at about 14 to 18 in. (35 to 45 cm) above the plants. Obviously this is much easier with the smaller growing globular cacti, rather than the columnar forms. However, even these kinds can be grown quite successfully too.

In addition to purchasing the lights a time-clock is more or less a necessity, as you must attempt to give the plants a day and night period. It is usually reckoned that the lights should be on for 16 to 18 hours for the spring to autumn period when plants are in growth, but cut back to about 14 hours for the winter.

Plants grown by this method without any daylight may need rather more personal attention, and care should be taken not to force the plants too much, as this can result in unnatural growth. In other words with the densely spined cacti a rather open structure would occur with areoles too far apart, and perhaps poor spination.

During the spring to autumn period high temperatures will be beneficial, but during the winter resting period, it is imperative that the temperature is kept reasonably low, preferably not higher than 45°F (7°C). If this is not done plants will continue to grow, and with most cacti a winter dormancy period is essential in order to ensure plenty of flower for the following year.

Seeds and the resulting seedlings can also be raised under artificial lights with equal success to those raised in a greenhouse. In fact given a higher winter temperature than we normally suggest they will grow continuously for at least the first two to two and a half years.

REVERSAL OF THE GROWING SEASONS

You will have gathered from earlier chapters that many plants in habitat have varying periods of the year when the rains come, often spring and late summer or winter. As a general rule plants from the USA can all be treated the same in cultivation, so that watering can be carried out from spring to autumn, and the plants left dry in winter for the resting period. If these same plants are grown in the southern hemisphere, it is possible for some species to continue growing during the same months of the year as they would in the northern hemisphere. However, most of these American desert plants will change their seasons to suit the climate occurring in the southern hemisphere. The same problems arise with plants native to the southern hemisphere, when introduced into the northern hemisphere. Some species will change their growing season to suit our climate, but not all of them.

PESTS AND DISEASES

Today there are many proprietary insecticides available from shops and horticultural firms which are suitable for use on all the plants mentioned in this book. Some kinds are contact sprays, others of a systemic nature. This latter kind means that the poison is absorbed into the plant, usually through the roots in such a way that sucking insects like mealy bug, scale insect and red spider soon kill themselves. For more information on cultivation and problems such as these we would refer the reader to our other book, *Pocket Encyclopaedia of Cacti in Colour*.

8. LOCATIONS VISITED
TEXAS

Location: This first location, a few miles to the north-west of Rio Grande City in Starr county is at an altitude of about 200 ft (61 m). Compared with other locations mentioned in this book—West Texas, Arizona and southern California—this is far from being a true desert area. The rainfall is quite high, about 28 in. (70 cm) per year, and because of this, most of the species of cacti growing here are quicker growers but shorter lived, and produce an abundance of seed each year, in order to ensure survival. This location can be divided into two sections; the low-lying part where the cacti are generally shorter lived; and the gravelly hillocks rising from it where two other species of cacti are to be found.

At certain periods of the year a great deal of rain can fall within a very short period of time, so that many of the smaller cacti may be submerged completely for a few hours. There can also be long periods of weather when the humidity is high. However, temperatures are rarely low so plant-life in general grows quite rapidly. Most of the smaller cacti probably reproduce from seed almost every year because warmth and moisture are available. In contrast the other localities studied in this book have an average rainfall of about 5 in. (12·5 cm) annually, and some years little or no rain at all. In those localities ideal conditions for seed germination may occur only once in seven years.

The countryside around here is fairly flat. The soil, which is of a clay type, bakes down very hard and cracks open in the heat following the rains. Various grasses grow well and the following species are likely to be found in such a location: *Ancistrocactus scheerii* (Fish-hook Cactus), *Coryphantha runyonii* (Big Nipple Cactus), *Dolichothele sphaerica*, *Echinocereus fitchii*, *Echinocereus melanocentrus*, *Echinocereus pentalophus* (Lady Finger Cactus), *Escobaria runyonii* (Junior Tom Thumb Cactus), *Thelocactus bicolor* var. *schottii* (Texas Pride) and *Homalocephala texensis* (Horse Crippler). This last species is really the largest growing globular cactus to be found in this part of Texas and can reach as much as 12 in. (30 cm) in diameter. Unlike other related cacti in what is often

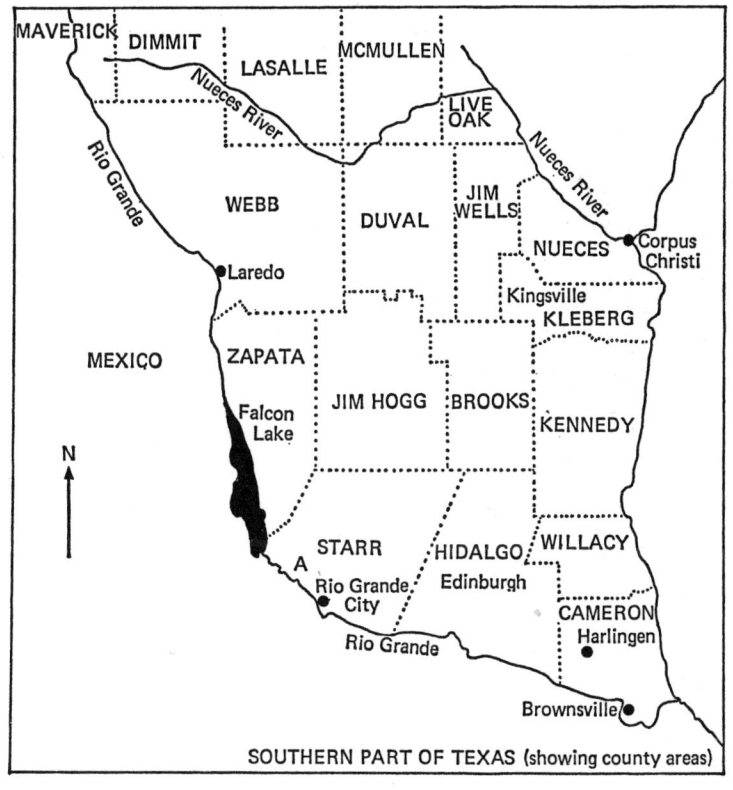

SOUTHERN PART OF TEXAS (showing county areas)

termed the Barrel Cactus Group, this species can stand far more water and has been found very close to the coast near Corpus Christi, which lies on the Gulf of Mexico. It has gained its common name of 'Horse Crippler' from the fact that the strong spines can easily penetrate a horse's hoof. This is likely to happen as it grows among grass, and an unwary horse can easily fail to see it. It is found over a very wide area of Texas and through into New Mexico and Oklahoma.

Other dwarf cacti also to be found growing in this region, sometimes among the grass but more often nearer bushes, were *Astrophytum asterias* (Sea-Urchin Cactus), *Echinocereus enneacanthus* (Pitaya) and *Mammillaria heyderi* var. *hemisphaerica* (Little Chilis). The first-named species is becoming increasingly difficult to find because of over-zealous collecting and the destruction of its native habitat for agriculture. The specimen illustrated (Plate 8) in habitat was 4 in. (10 cm) in diameter, quite a fair size, but specimens 6 in. (15 cm) in diameter have been found. Without any spine covering to protect it from the sun, a cactus such as this would soon die under full sun conditions. It is a very interesting species as in periods of drought it shrinks and tends to pull itself down level with the surrounding ground, so that any loose soil can cover it. In this respect it is similar to *Ariocarpus fissuratus* (Living Rock Cactus), which you can see illustrated in habitat later (Plate 37).

Three species of *Opuntia* (Prickly Pear) which were also seen growing here are also to be found over a very wide area in the USA and Mexico. These were a miniature crawling species *Opuntia schottii* (Dog Cholla or Devil Cactus) a thin cylindrical stemmed species forming into low bushes, *Opuntia leptocaulis* (Tasajillo or Desert Christmas Cactus) and a flat padded species, *Opuntia engelmannii* (Engelmann Prickly Pear). The last species is illustrated in Plates 31 and 32.

A few other succulent plants were also to be found here including *Jatropha berlandieri*, a member of the Euphorbia family. This plant has quite a large tuberous root, while its short erect aerial growth consists of a few pale blue leaves and clusters of small red or orange red flowers. It is also possible to find scattered specimens of the genus *Manfreda*. There are a number of species. The plants consist of rosettes of soft fleshy leaves 12 in. (30 cm) or more in diameter. The leaves themselves vary in colour from green to bronze red, with irregular markings on them. As the leaves are so soft and fleshy, grazing animals will eat them. In order to survive nature has provided this plant with almost an underground bulb-like base and a very succulent swollen root structure, so that this plant can regrow if the leaves have been eaten or died off in time of prolonged drought. *Manfreda* is related to *Agave* (Century Plant), but its flower spikes rarely exceed 5 ft (1·6 m) in height compared with those of the larger-growing *Agave* species which often exceed 20 ft (6·1 m) in height—rather like telegraph poles! On rising ground in this area, where the soil was poorer and much more gravelly, two other species of cacti were found growing among dense stands of *Prosopis juliflora* (Mesquite). One of them, *Wilcoxia poselgeri* (Dahlia Cactus), depends for support on the thorny stems of the Mesquite on which it can rest its slender and quite delicate stems. *Wilcoxia poselgeri* is aptly named Dahlia Cactus because of its subterranean root structure which consists of a cluster of tuberous roots. This is another plant which can re-grow from below ground if the aerial part of the plant is killed or dies away during drought. The slender stems grow quite quickly with the right climatic conditions and its attractive flowers are quickly produced. It can grow to 2–3 ft (60 cm–91 cm) among Mesquite, but if found in more open conditions rarely grows taller than 12 in. (30 cm).

The other species which grows here in the shade of the Mesquite is *Lophophora williamsii* (Peyote or Dry Whisky), very well known through its alkaloid content, which has made it illegal for anyone to collect and transport it in the USA. Many articles and a book have been written about this plant and its connection with certain religious festivals of the Indians. It sometimes grows in clusters. The plant is small and spherical with a chalky-blue body, but completely spineless, with only small tufts of white hair. It possesses a lengthy tap root, rather like a carrot. The flowers can vary from white to yellow or pink; the latter colour is the most common. It is a wonderful little species which produces its small flowers over a very long period. The filaments of the stamens are so sensitive to touch that the stamens will close over the stigma. Nature's idea is that when insects have touched the filaments of the stamens (see diagram, page 229) they will be unable to fly away without taking some pollen with them. With any luck they will take this pollen to another flower and so aid pollination, which in turn will result in seeds and ensure the continuation of the species. This is not the only plant to possess this interesting feature; in fact most *Opuntias* do. You can test it by putting your finger into the flower of an *Opuntia*, and within a moment or so the stamens will start to close inwards.

PLATES

1 View across the 6-ft (2-m) high bushes of Mesquite a few miles N.W. of Rio Grande City, Starr county, Texas.
2 *Homalocephala texensis* in habitat.
3 Close-up of the flowers of *Homalocephala texensis*.
4 Cluster of the 1-in. (2·5-cm) high flowering heads of *Escobaria runyonii*.
5 *Ancistrocactus scheerii*.
6 *Dolichothele sphaerica*.
7 *Thelocactus bicolor*.
8 *Astrophytum asterias* in habitat.
9 Habitat of *Lophophora williamsii*, in which three specimens can be seen.
10 *Astrophytum asterias* in flower.
11 Close-up of the ¾-in. (2-cm) diameter flower of *Lophophora williamsii*.
12 *Opuntia schottii* with the flattened central spines being clearly visible.
13 Flowering stems of *Wilcoxia poselgeri*. Note the variation of the size of the flowers. In fact the flower gets larger the longer it stays open.

LOCATION B: (PLATES: Nos. 14 to 25, MAP: p. 41)

Location: The next area is about thirty or so miles (48 km) south-east of Marathon, Brewster county in what is called the Mesa country. This location

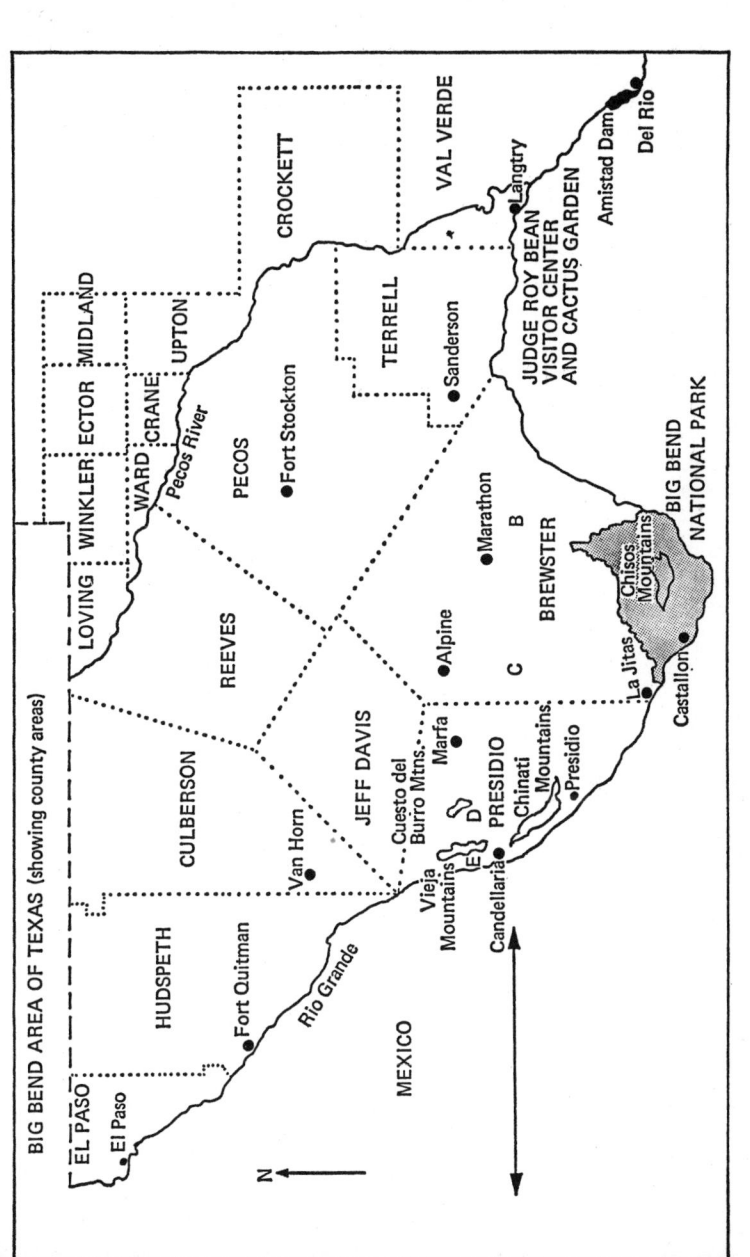

BIG BEND AREA OF TEXAS (showing county areas)

we visited is the south-western part of this area, and has numerous hills with flattened tops (hence *Mesa*, from *mensa* Latin for table) which you can see in the background (Plate 14). A very long time ago, when this area was under water, erosion occurred leaving behind the hard limestone, which is what you can see today. Except in some of the valleys where soil has been washed down over the years, plant life has quite a job to survive here, often just growing from between cracks in the rocks, or among partially broken-down limestone.

Quite a variety of cacti were to be found, also various grasses including *Buchloe dactyloides* (Buffalo Grass) which grew where there was a greater soil depth at the bottoms of slopes. In these same places we also found a few low-growing Juniper trees along with *Prosopis glandulosa* (Honey Mesquite), stunted specimens of *Flourensia cernua* (Tar bush) and the *Larrea tridentata* (Creosote bush). These grew sparsely over a very wide area here, and also in other parts of the Big Bend area of Texas. Many smaller cacti grew beneath these trees, for the seedlings would never survive without their shade. Most cacti produce an abundance of seed in their fruits and in areas such as these where moisture is minimal, it may be many years before suitable conditions occur for the seeds to germinate. Even then the number of seedlings which survive is very small. This is ranch land; the ranchers reckon on one cow per fifty acres (20 hectares), which give you some idea of the quality of the grazing! These cacti are mostly long-lived in contrast with those species shown earlier growing in the south-eastern part of Texas, nearer Rio Grande city. *Echinocereus dasyacanthus* (Rainbow Cactus) was quite common here, mostly as single or double headed specimens. The plants we examined had grey to creamy white coloured spines. Illustrations of this species appear later along with red spined forms (see Plates 46 and 47 taken in the Cuesto del Burro Mountains). Scattered specimens of *Coryphantha echinus* (Hedgehog Cory-Cactus) were also to be found, and *Euphorbia antisyphilitica* (Candellaria), both also to be found blending in with the dirty-white coloured limestone to such an extent that if you did not mark the spot you could hunt for ages before finding them again.

In some books you will also find the name *Coryphantha pectinata*, but this is only the juvenile form of *C. echinus*. Although it is the juvenile form without any prominent central spines, it is quite capable of flowering as you can see in Plate 16. Some variation occurs as regards the flower structure such as petal shape, colour, but basically the flowers can open far wider with juvenile specimens, as there are no erect central spines to stop this.

In one spot where the soil conditions were much better we came across a colony of *Echinocereus triglochidiatus* var. *octacanthus* (Claret Cup Hedgehog) in bud. Compared with the type species *E. triglochidiatus* it has very robust stems and grows much taller. Another interesting feature with the flowers is that they stay open day and night, whereas most cacti flowers only open in daylight hours.

The most spectacular tree in the Mesa country is *Yucca thompsoniana* (Thompson Yucca or Bayonet Plant). It is a slow-growing species with a main stem and branching above ground in much the same way as its better known relative *Yucca brevifolia* (Joshua Tree). The waxy flowers of these Yuccas are most attractive and highly scented. With the aid of the Yucca moth, which helps with the pollination, they seed freely. This species of Yucca is very popular for landscaping.

Another common desert plant in these parts is *Dasylirion leiophyllum* (Sotol). It is more or less a stemless plant averaging 3 ft–4 ft (91 cm–1·3 m), high with narrow leaves and teeth along the margins. It seems capable of growing under very dry conditions, but equally enjoys plenty of water when grown in cultivation. From this plant the Mexicans make a very potent alcoholic brew which is called Sotol, and looks rather like dirty washing-up water. This drink has quite a kick to it, and is comparable in strength (or possibly stronger) to another better-known beverage from Mexico, Tequila which is distilled from the juice of Agaves. There are a number of species of *Dasylirion* ranging over the western half of Texas, through Arizona and over a wide area of Mexico.

In this Mesa country we frequently came across another xerophytic bush *Koeberlinia spinosa* (Christ's Crown). This is a green thorny bush without leaves, which in these parts rarely exceeded 18 in. (45 cm) in height, but can grow to as high as 6 ft (1·8 m). When we came across these low thorny masses, it was always worth an investigation as to what we might find growing beneath it. On more than one occasion we came across *Hamatocactus uncinatus* (Cat-Claw Cactus) growing in the protection of this bush— although it is seeking a little shade rather than protection. It is a species of cactus which has been listed under various genera including *Thelocactus*, *Ferocactus* and *Glandulicactus*, or is just referred to as *Echinocactus uncinatus*. The flowers vary in colour, but are never bright, in contrast to the wonderful red fruits which appear later. The hooked central spines can be very long, as with the specimen illustrated (Plate 22) where they are almost 6 in. (15 cm) long. This species never seems to grow in colonies, only scattered specimens are found. The other small cactus growing here was *Neolloydia conoidea*, also known as *N. texensis*. It is a bluish-bodied plant, with white radial spines and black centrals, which produces over a period of a few weeks many attractive magenta blossoms. In the driest locations the plants rarely had more than one or two heads, but we came across one small area where we found large clustering plants, some with over fifty heads. These were growing in much better soil, and this proves, as would be expected, that when a species finds a good location, much better specimens result. It has often been stated that in cultivation one should imitate the habitat conditions regarding soil, etc., but this is not strictly true, as plants such as cacti do not always select the best site.

PLATES

14 *Yucca thompsoniana* growing in the Mesa country some thirty or so miles (48 km) south-east of Marathon, Brewster county, Texas.

15 One specimen of *Coryphantha echinus* is visible nearer the largest stone at top right of picture.

16 Two specimens of *Coryphantha echinus*. The right-hand specimen is the juvenile form, which has in the past been known under the name *C. pectinata*.

17 A number of plants of *Dasylirion leiophyllum* in habitat some 30 miles (48 km) south-east of Marathon, Brewster county, Texas.

18 *Echinocereus triglochidiatus* var. *octacanthus* showing an 18-in. (46-cm) diameter plant.

19 Close-up of top of stem and flower of *Echinocereus triglochidiatus* var. *octacanthus*.

20 A specimen of *Hamatocactus uncinatus* is just visible in the midst of this xerophytic bush (*Koeberlinia spinosa*) some 30 or so miles (48 km) south-east of Marathon, Brewster county, Texas.

21 Close-up of flower of *Hamatocactus uncinatus*.

22 Very long central spined form of *Hamatocactus uncinatus*.

23 Flowering-plant of *Neolloydia conoidea* growing amidst another bush of *Koeberlinia spinosa*.

24 Fine clustered specimen of *Neolloydia conoidea* at this locality.

25 Close-up of *Neolloydia conoidea*.

LOCATION C (PLATES: Nos. 26 to 42, MAP: p. 41)

Location: This site is about sixty miles (96 km) south of Alpine in Brewster county, and it proved to be very interesting indeed. It could be divided into two areas. This ranch land was desert-scrub with a somewhat sandy soil providing little good grazing for cattle. Rising out of this desert-scrub were numerous limestone outcrops and ridges on which we found a variety of cacti. In the valleys or low lying areas, which you can see at the top left of Plate 26, two other species of cacti occurred which were not found on the rocky outcrops.

On these rocky outcrops and on the slopes of the rocky ridges the chief vegetation consisted of 3-ft to 4-ft (91-cm to 1·3-m) high specimens of *Larrea divaricata* (Creosote bush) and scattered specimens of *Agave lophantha* var. *poselgeri* (Lechuguilla), a dwarf but quite viciously-armed species which in some areas grows in abundance, making walking or climbing a very hazardous operation. On the slopes here we found many huge clumps of *Echinocereus stramineus* (Strawberry Cactus or Pitaya) in full bloom. They form into large mounds made up of a hundred or more stems, and can exceed 3 ft (91 cm) or more in diameter. This is a very beautiful species with a wide distribution over Texas from the Pecos River in the east to El Paso in the west, just into New Mexico and over a wide area of the northern part of Chihuahua in Mexico which is to the south. Some variation does occur within the species, although compared with other species in the same genus it does appear to

D

be more stable, in that specimens within one colony show little variation. On these slopes the only young specimens of this *Echinocereus* to be found were growing in the shade of the Creosote bushes, along with the occasional specimen of *Escobaria tuberculosa* (Cob Cactus). This is a small cylindrical cactus which branches sparingly and rarely exceeds 6 in. (15 cm) in height. It is densely covered with grey to brown coloured spines. On old plants the spine clusters are usually missing, leaving just the bare tubercles—hence its common name Cob Cactus. The small 1-in. (2·5-cm) diameter pink flowers appear at the apex of each stem. This species was growing in great abundance between clefts of rock on the rocky outcrops nearby. We found five species of Opuntia on these same slopes, two flat padded kinds *Opuntia engelmannii* (Engelmann Prickly Pear) and *Opuntia macrocentra* (Purple Cactus or Long spined Prickly Pear). Quite large clumping specimens of these species were scattered over this area. *O. engelmannii* was by far the most common. Orange-flowered plants of this species occurred occasionally, although yellow was found more frequently. *O. macrocentra* is a very attractive species because of the pale blue to purplish coloured pads, while the yellow flowers often have reddish centres. Two species of Cholla Opuntia were also scattered around, but with the poor soil and accompanying dry conditions the specimens of *O. leptocaulis* (Tasajillo or Desert Christmas Cactus) rarely exceeded 2 ft (60 cm) in height, whereas those of *O. imbricata* (Cholla or Cane Cactus) attained at least 4 ft–5 ft (1·3 m–1·6 m). The first of these two species is usually densely branched with very slim stems or joints, and long sheathed spines which are viciously barbed. You cannot see the barbs with the naked eye, but you soon become aware of their presence if you get too near! The flowers are green or yellow, and they are followed by attractive red fruits in the winter. *O. imbricata* is dealt with in more detail later (Page 50 and Plate 45).

The fifth species of Opuntia here was a dwarf mat-forming one, called *Opuntia schottii* var. *grahamii* (Graham Dog Cactus or Mounded Dwarf Cholla), which grows over a wide area of Texas and further west. It rarely exceeds 4 in. (10 cm) in height, and grows flat to the ground. When in sandy soil it is sometimes partly buried. The yellow flowers are 2 in. (5 cm) in diameter. The spines are as vicious as those of *O. leptocaulis*, and as the joints are easily detached under dry weather conditions, they can be transported by attaching themselves to passing animals, only to drop off many miles away and grow into a new plant. We noticed some clumps of this species which were apparently quite dead owing to drought, but on digging one up we found that the swollen root was still alive. So once conditions were right this old root would be quite capable of growing new joints and forming a new carpet of this dwarf viciously-spined species.

Our attention was next concentrated on a few of the highest rocky outcrops in the vicinity, where we found not only many more specimens of *Escobaria tuberculosa* and *Echinocereus stramineus*, but also two species of Mammillaria. These were *M. pottsii* (Desert Rat-Tail Cactus) and *M. meiacantha* (Heyder Nipple Cactus). Compared with the other species we had seen, these were quite rare. *M. pottsii* is a sparsely clustering species with 6 in. (15 cm) cylindrical stems, densely covered with short brown- or black-tipped spines. The flowers are reddish-brown and appear on the sides of the stems nearer the apex. *M. meiacantha* is a flattened globular species which grows flush with soil, and is rather similar in general appearance to *M. heyderi* var. *macdougalii* (Plates 60 and 61).

We then moved down to the valley area where the soil was partly fine sand or baked clay, and the specimens of *Flourensia cernua* (Tar bush) and *Larrea divaricata* (Creosote bush) were nearly 6 ft (2 m) in height. Here we found a species of Yucca, or *Y. torreyi* (Spanish Dagger), mostly around 10 ft (3·3 m) in height. This is one of the species of Yuccas where the fibres in their tough leaves have been extracted and made use of, in much the same way as with Agaves, from whose fibres sisal string is made. *Y. torreyi* is not common when the Creosote Bush is in abundance, or where there is plenty of grass. It seems to prefer locations where these other types of vegetation have disappeared through excessive grazing in the past.

It was in this same area that we saw two other cacti for the first time—*Echinocactus horizonthalonius* (Blue Barrel or Eagle's Claw) and *Ariocarpus fissuratus* (Living Rock). Scattered plants of these two species were growing here, with those of *E. horizonthalonius*, either in the open or sometimes beneath a Creosote Bush. Normally this species is solitary, but we did find a five-headed specimen beneath one such bush. These plants are either flattened globular or short cylindrical in form, but the spination can vary considerably from one plant to another in the same locality. Sometimes the cylindrical specimens with a slightly spiral rib structure are far more densely spined; this form is known by some people under the variety name of *spinescens*. The flowers can also vary a little in colour and actual petal shape.

Ariocarpus fissuratus is a most amazing species with a completely flattened top. It usually grows with this top level with the surrounding ground, in this case in full sun and in a clay-type of soil. Without the human hand shown in Plate 37 it would require a very close look at this picture to find the plant. Flowers generally appear in the autumn and in nature this is the only time this species gives the game away as to where it is hiding! Like its relatives in Mexico, it is a very slow-growing species, and specimens only a few inches across could be a hundred years old or more.

Not many miles south from this location we again found *Ariocarpus fissuratus*, *Echinocactus horizonthalonius*, *Mammillaria meiacantha*, *M. pottsii* along with two other miniature cacti *Epithelantha bokei* (Button Cactus) and *Mammillaria lasiacantha*. (Lace-spine Cactus). Single heads of either of these miniature cacti rarely exceed 1 in. (2·5 cm) and usually much less. *E. bokei* is usually a solitary species, but has much larger flowers than the better-known *E. micromeris* (Plate 41). It is shown growing with *Selaginella lepidophylla* (Resurrection Plant), and in one photo it is in flower, whereas the *Selaginella* is rolled up and looking quite dead; when it is in growth the *Selaginella* opens like a rose and is bright green. There are in fact a number of species of these xerophytes. It would appear that they are one of the plants which help with the propagation of certain miniature cacti. The seeds of these miniature cacti may lodge in one of these Resurrection Plants and when rains come they are able to germinate under their shade, whereas in full sun conditions seedling cacti would not live very long. Beneath many of these Resurrection Plants we also found numerous seedlings of *Epithelantha bokei* and *Mammillaria lasiacantha*. In general appearance these two cacti look rather alike if they are not in full bloom, but once you know a little bit more about cacti you can easily distinguish them even when they are not in bloom. Many seedlings of these same species could easily be found in rock crevices where sufficient shade could be obtained. It is interesting how in nature there are many instances where one species of plant life can help another.

PLATES

26 View some 60 miles (96 km) south of Alpine, Brewster county, Texas.

27 3-ft (91-cm) diameter clump of *Echinocereus stramineus* in bloom. Note one small specimen of *Escobaria tuberculosa* close to the right of the *Echinocereus*, whilst on the extreme right in the background you can see the thin but very spiny stems of *Opuntia leptocaulis*.

28 One of the authors (Brian Lamb) measuring a specimen of *Echinocereus stramineus* at the same locality.

29 Close-up of the flowers of *Echinocereus stramineus*.

30 Close-up of the fruit of *Echinocereus stramineus*.

31 4-ft (1·3-m) diameter clump of *Opuntia engelmannii*.

32 Orange-flowered form of *Opuntia engelmannii*.

33 3-ft (91-cm) diameter mass of *Opuntia schottii* var. *grahamii*.

34 5-in. (12·5-cm) diameter specimen of *Echinocactus horizonthalonius* growing in semi-shade beneath a Creosote Bush.

35 *Yucca torreyi* with old flower remains in the valley area 60 miles (96 km) south of Alpine, Brewster county, Texas.

36 Close-up of *Echinocactus horizonthalonius* in bloom.

37 *Ariocarpus fissuratus* in habitat at the same location as *Y. torreyi*.

38 Close-up of *Ariocarpus fissuratus* in bloom.

39 *Epithelantha bokei* in flower, surrounded by a dormant specimen of *Selaginella lepidophylla*.

40 Close-up of *Epithelantha bokei* in bloom; flowers can vary from white to pale pink.

41 *Epithelantha micromeris* in bloom, from Carlsbad, New Mexico.

Note. Specially illustrated here so that you can see it along-side *E. bokei*.

42 Close-up of the top of one stem of *Mammillaria pottsii* from 60 miles (96 km) south of Alpine, Texas.

LOCATION D (PLATES: Nos. 43 to 48, MAP, p. 41)

Location: Just beyond Marfa in a south-westerly direction the road peters out, and then we were on a dirt road which steadily got worse the further we went. Around us were quite green rolling hillsides at an altitude of around 4,000 ft (1,300 m). These hillsides were fairly well covered with grasses, making good grazing land; these included 'Chino Gramma' grass, and many patches of *Nolina erumpens* (Mesa Sacahuista or Basketgrass). Unlike its better-known relatives *N. recurvata* and *N. longifolia* from Mexico, this is a stemless species, but the foliage is quite similar, with rather rough leaf edges, like Pampas Grass. The inflorescence is creamy-yellow and also similar to its relatives in general appearance unless examined closely. Further on we started climbing through the Cuesto del Burro mountains to our next location, where the scenery was very spectacular.

Here the vegetation thinned out considerably, but being at a somewhat higher altitude than earlier on, we found some very fine specimens of *Fouquieria splendens* (Ocotillo) in full bloom. These were flowering at least three weeks later than those at a lower altitude. They were 15 ft (5 m) in height, whereas those seen in the Mesa country were usually quite stunted. A wax has been prepared from the juice of the stems, whilst the Apache Indians have used them medicinally in a number of ways, including the relief of tiredness by washing in the juice. Plates 43 and 44 will give you a good idea of this xerophytic shrub, which is found over a wide area of western Texas, southern New Mexico, Arizona and into Mexico.

We also found growing among the Ocotillo scattered specimens of *Opuntia imbricata* (Cholla or Cane Cactus), and these were in bud. Just a little further south on the much drier and hotter south-facing slopes we found them again, in bloom. Most specimens here averaged around 6 ft (2 m) in height. The fruits of this species when ripe are most unusual, being bright yellow and very corrugated in structure.

Beneath these Opuntias and in the open among dried grass or

growing with *Agave lophantha* var. *poselgeri* (Lechuguilla), were numerous clustering specimens of *Echinocereus dasyacanthus* (Rainbow cactus), some with very reddish-coloured spines. The flower colour varies from pale yellow to a rich buttercup-yellow, but always with a green centre. Occasionally pink-flowered forms have been found, or even a plant producing yellow flowers from one stem and pink from another. The flowers can be as much as 6 in. (15 cm) in diameter. Plates 46 and 47 give you a good idea of this beautiful species.

We also found a few poor specimens of *Echinocereus dubius* (Purple Pitaya), growing in association with *Echinocereus stramineus* (Strawberry Cactus or Pitaya). Natural hybrids in the wild amongst cacti are quite rare, but one specimen was found here, almost intermediate between these two species of *Echinocereus*. It had long sprawling stems as with many forms of *E. dubius*, and had a similar number of spines, but these were much stronger. It had the same flowers as *E. stramineus*.

At this same location we again came across *Euphorbia antisyphilitica* (Candelilla or Wax Plant) which we had first seen back in the Mesa country, but these were much finer specimens. They were densely clustering and many were in bloom. The flowers are very small, produced on the upper sections of the stems, and rather star-like in appearance with white petals and a pink centre. The plants are harvested by the Mexicans, and the wax obtained has a number of uses.

Nearby also growing with *Euphorbia antisyphilitica* we found a very fine specimen of *Hamatocactus hamatacanthus* (Giant Fish-Hook Cactus), shown in Plate 48. We never found colonies of this plant, as with *Mammillaria pottsii* mentioned earlier on in the book, just scattered specimens. It can grow to a foot or so in height (30 cm) upwards, and there is considerable variation in spine count, structure and length. Some forms, for example, can have very long hooked central spines. The flowers vary from pale yellow to a rich buttercup-yellow, with many of the outer petals tinged with red. Plants are usually solitary but clumping plants are sometimes found. Its distribution is widespread over the Big Bend area of Texas.

PLATES

43 General view in the Cuesto del Burro mountains, with flowering stems of *Fouquieria splendens* in the foreground.

44 Close-up of the top 3 ft (91 cm) of *Fouquieria splendens*.

45 6-ft (2-m) high specimen of *Opuntia imbricata* in bloom, on the southerly slopes of the Cuesto del Burro mountains.

46 Reddish spined form of *Echinocereus dasyacanthus* growing in association with *Agave lophantha* var. *poselgeri* in the Cuesto del Burro mountains.

47 Very old specimen of *Echinocereus dasyacanthus* from the same locality. This plant must be a great age.

48 Good specimen of *Hamatocactus hamatacanthus* growing beneath a Mesquite tree and surrounded by *Euphorbia antisyphilitica* in the Cuesto del Burro mountains.

LOCATION E (PLATES: Nos. 49 to 54, MAP, p. 41)

Location: Down by the Rio Grande, which is the border between Texas and Mexico, a few miles north of Candellaria. Candellaria consists of just a few rough dwellings and a small trading store. The dirt road is very rough indeed, and the last few miles to this location are very tough on any vehicle, which is why few people have visited this part.

Although there is little water by the Rio Grande, the vegetation is very dense, with masses of *Tamarix gallica* (Saltcedar), a considerable quantity of *Prosopis juliflora* (Mesquite) and a few willows. It was among the Mesquite, a rather thorny bush, that we found *Peniocereus greggii*. There are a number of common names for this cactus including Deerhorn Cactus and Queen of the Night, but this latter name is also used for other night-flowering cacti. *Peniocereus greggii* is a very unusual cactus in that it has huge tubers, which can vary in weight from 1 lb or so to 30 or 40 lb (13·6 kg–18·1 kg). The local Indians and Mexicans eat it, and it is said to be similar in taste to turnip. They also use it for medicinal purposes, placing a large piece on one's chest, but although they believe in it, it is unlikely to be much help for chest ailments.

These tubers bear a number of thin stems, usually 4 or 5 ribbed, which need support because they are so thin. The best specimens are always found among dense bushes where there is shade, and also where the stems can obtain support. The flowers appear in the spring or early summer. They are highly scented, but last only one night. Three illustrations are shown of this species, including detailed photos of the root, flower and fruit.

Nearby we also found *Echinocereus dubius* (Purple Pitaya or Alicoche) in huge sprawling masses, growing in very sandy soil in odd spots among the dense bush where the sun could get through. This species was also seen by us in much drier locations, but there the stems tended to be much shorter. Plate 53 shows a mass of *Echinocereus dubius* to the right of the picture; a plant of *Peniocereus greggii* was found under the tree to the left. A close-up of *Echinocereus dubius* is also shown in flower (Plate 54).

Only a mile or so from this locality on higher ground among bare rocks and so in a much drier position, we came across *Mammillaria microcarpa* (Fish-Hook Cactus or Lizard Catcher). Illustrations of those in Arizona are shown in Plates 79 and 94. The only difference with the specimens from the Texas/Mexico border is that they were flattened globular plants and few tended to become columnar; also they were not quite so densely spined, so that the body of the plant could be seen.

Nearby were quite large clusters of *Epithelantha micromeris* (Button Cactus). A close-up of a single-headed specimen is shown on Plate 41. These wonderful plants are slow-growing. Clumps of a few hundred heads are not at all uncommon, each head rarely being over 1 in. (2·5 cm) in diameter. It is more than likely that some of the large clusters could be over a hundred years old at least! Also growing from among rocks we found *Opuntia rufida* (Blind Prickly Pear), a species very well known to cactus enthusiasts. The flat pads are rarely more than 6 in. (15 cm) in diameter, often much less, and covered at intervals with clumps of brown glochids. If touched these stick into you as they are barbed, but they are not poisonous. This species has yellow flowers about $1\frac{1}{2}$ in. (3·8 cm) in diameter (Plate 139).

East and south of Candellaria, not far away from here on clay hills which are slowly eroding, the remains of forests, now fossilised, can be found.

PLATES

49   View of the Rio Grande, what there is of it at this point on the Texas/Mexico border, a few miles north of Candellaria. Note how dense the vegetation is here, due to the nearness to water. Less than 50 yards (46 m) from this spot to the right we found specimens of *Peniocereus greggii*.

50   Large tuber of *Peniocereus greggii* visible to the right of the pick handle.

51   Flowers of *Peniocereus greggii*.

52   One stem and fruit of *Peniocereus greggii*.

53   View showing mass of *Echinocereus dubius* to the right of the picture, whilst a specimen of *Peniocereus greggii* was found just to the right of the Mesquite bush on left of the picture.

54   Top of stem and flower of *Echinocereus dubius*.

ARIZONA

LOCATION F (PLATES: Nos. 55 to 61, MAP, p. 56)

Location: About seventy miles (113 km) west of Nogales, on the Arizona/ Mexico border, and just a few miles south of Arivaca in Pima county. This was rolling hillside country, covered with grass. Even in the heat of early summer this was still green on the north-facing slopes, whereas on the slopes facing the sun the grass had turned to golden-brown. The soil was fine, very red in colour, and the smaller cacti were mainly to be found on or near the top of hillocks often where there were rocky outcrops. The altitude here was around 3,300 ft (1,000 m).

Many of the slopes were in parts quite densely covered with *Fouquieria splendens* (Ocotillo), and scattered specimens of *Ferocactus wislizenii* (Candy or Fish-hook Barrel), some of them 3 ft–4 ft (91 cm–1·3 m) in height.

On one such hillock four of the smaller cacti were to be found, growing among the grass and only visible because of their flowers. The scattered specimens of *Echinocereus rigidissimus* (Arizona Rainbow Hedgehog), could easily have been missed except for their flowers as you can imagine by looking at Plate 56. This particular species, as with the other closely related species *E. dasyacanthus*, can also be quite variable in spine colour, and that is why it is commonly named Arizona Rainbow Hedgehog. You can see from Plate 57 how red some forms can be. In one specimen you can get changes of colour in rings all the way up a stem.

In habitat *Mammillaria heyderi* var. *macdougalii* (Cream Cactus) grows with its top surface almost flush with the surrounding soil as in Plate 60. It is sometimes to be found beneath deciduous trees where the plants are almost covered with dead leaves, making them invisible. However, when the rains come they do plump up a bit, and tend to look more like the one shown in Plate 61. This species can grow up to around 8 in. (20 cm) in diameter and has a carrot-like tap-root. The cacti so far mentioned were far from common, but *Coryphantha vivipara* var. *bisbeeana* (better known under *Coryphantha aggregata*) and *Mammillaria oliviae* were few and far between. The first of these forms into large clusters eventually and has the common name of Hen and Chicken Cactus. It has tapering petals and a paler flower colour than its relative *C. vivipara* var. *arizonica* (Plate 107). The other species to be found here was a much smaller Mammillaria, *M. oliviae* (Olive's Pincushion) which is not illustrated. This species is a small globular or short cylindric type, almost completely white spined, but some specimens have brown tips.

On nearby slopes, often growing in association with the Ocotillo, clumps of *Agave parviflora* were to be found. This is a beautiful little species whose single heads rarely exceed 6 in. (15 cm) in diameter. The leaves are narrow, green with white markings and similar coloured hairs along the margins. As the name suggests, the flower and flower-spike are small compared with some of its large-growing relatives such as the blue-leaved *Agave palmeri*, also to be found in many parts of this section of Arizona (Plate 97).

PLATES

55 View a few miles south of Arivaca and some seventy miles (113 km) west of Nogales, in Pima county, Arizona. To the left of centre can be seen a specimen of *Ferocactus wislizenii*, while to the right and in the foreground are stems of *Fouquieria splendens* (Ocotillo).

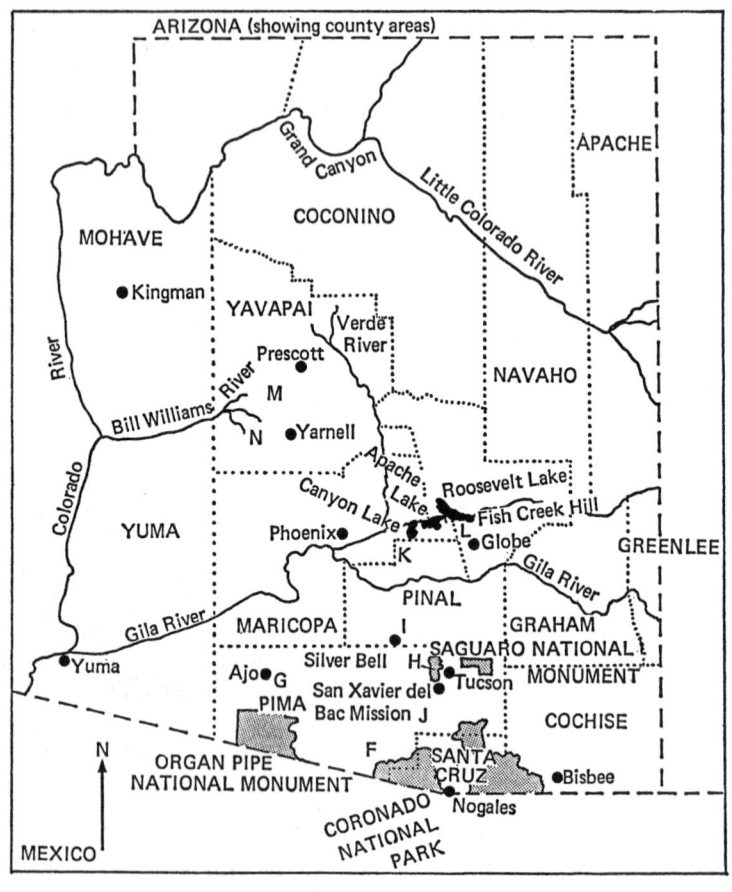

ARIZONA (showing county areas)

56 Closer view of this same area, where it is possible to see four specimens of *E. rigidissimus*, one of which is in flower.

57 Fine specimens of *Echinocereus rigidissimus* in bloom, shown in habitat. It measures just over 6 in. (15 cm) in height.

58 Close-up of the flowers of *Echinocereus rigidissimus*.

59 Close-up of the top of a red spined form of *E. rigidissimus* with one flower bud just forming.

60 *Mammillaria heyderi* var. *macdougalii* in natural surroundings.

61 Close-up of a 6-in. (15-cm) diameter specimen of *M. heyderi* var. *macdougalii*.

LOCATION G (PLATES: Nos. 62 to 66, MAP, p. 56)

Location: These pictures were filmed some three miles (5 km) south of Ajo, in Pima county, south-western Arizona, at a place a few miles north of the entrance to the Organ Pipe National Monument. The Monument is an area of some 500 square miles (12,950 hectares) of beautiful Sonoran desert scenery, in which a variety of cacti are to be found. The name Organ Pipe is in fact the common name of another spectacular Arizonan cactus, *Lemaireocereus thurberi*.

At this spot the country was slightly undulating. The ground varied from sandy desert type to, more often, a compacted gravelly soil, which is probably why the specimens of *Carnegiea gigantea* were far from being robust. In fact most of them were younger single stemmed specimens.

This is the most northerly spot at which *Lemaireocereus thurberi* is found, as it is really a Mexican species which has encroached into the USA. It would appear that these specimens are more branched than those further south, sometimes because frost damages the growing tip, which means that damaged stems produce one, two or three branches from beneath the damaged part, or even from ground level.

The golden-yellow spined form of *Echinocereus engelmannii* was also common here, and this goes under the varietal name *nicholii* (Nichol's Hedgehog Cactus). The specimen growing in the Silver Bell location nearer Tucson is illustrated in Plate 76. Scattered specimens of *Ferocactus covillei* (Coville's Barrel Cactus) were also to be found here, ranging from 2 ft to 3 ft (60 cm to 91 cm) in height. Even this species reaches a height of 5 ft (1·6 m) on occasions. It is a very strongly spined species, with one hooked central which can be as much as 5 in. (12·5 cm) in length. The specimen shown in Plate 65 was a fairly young specimen, but absolutely perfect, about 18 in. (46 cm) in height and diameter. Young specimens of any of these species of *Ferocacti* are tubercled in appearance, that is they do not have a distinct rib structure. With most species around ten years of age, the tubercles start to unite to form this rib structure; this also occurs with many of the closely related *Echinocacti*. A few of the Cholla Opuntias were also observed here, including *O. fulgida*, *O. arbuscula* and *O. spinosior*, but specimens tended to be rather stunted owing to the poor soil.

PLATES
62   A fine multi-branched specimen of *Lemaireocereus thurberi*, being admired by our friend Don Johnson. This was just south of Ajo in south-west Arizona.
63   Close-up of a 2½-in. (6-cm) diameter flower of *Lemaireocereus thurberi*.
64   Closer view of a fine specimen of *Lemaireocereus thurberi* which was not far off 20 ft (6·1 m) in height. Note the flower buds just forming near the tips of some of the stems.
65   18-in. (46-cm) diameter specimen of *Ferocactus covillei*.
66   Close-up of the flowers of *Ferocactus covillei* on a slightly smaller specimen.

LOCATION H (PLATES: Nos. 67 to 72, MAP, p. 56)

Location: The Saguaro National Monument which consists of over 120 square miles (31,080 hectares), is in two sections, east and west of Tucson, in Pima county, Arizona. These photographs were taken in the western section of the Tucson Mountain Park, some 16 miles (26 km) west of Tucson.

In many parts of the southern half of Arizona *Carnegiea gigantea* (Saguaro) is a very prominent feature of the landscape, and we feel it should be given a section of its own, more or less ignoring the other species of plants, particularly cacti, which grow there. The flower of *Carnegiea gigantea* is the state flower of Arizona and was featured on a USA postage stamp in 1962.

Average good specimens of this giant Saguaro cactus are around 30 ft–40 ft (9·1 m–13·3 m) in height, but occasional specimens nearer 50 ft (15·2 m) have been known. Their main stem diameter can be as much as 2 ft 6 in. (76 cm). They can be found in various habitats, in rocky locations or in sandy, more desert-like situations as here. Where the rainfall is less most specimens are then to be found following the line of washes. They have been found at altitudes ranging from around 600 ft (183 m) up to over 3,000 ft (1,000 m) and in the more northerly locations in central Arizona, only appear on the southern slopes of hills. The rainfall here can vary from as little as 3 in. (7·5 cm) to 11 in. (28 cm) per year.

As you can see in Plates 67 and 68 these huge cacti branch, but not until they have reached perhaps 15 feet (5 metres) in height. The root system of these cacti is usually rather shallow and can be compared to the spokes of a wheel, radiating out in all directions. This is certainly true of specimens growing in level sandy-soil conditions, where they have to be capable of making use of water from the lightest rainfall. In wet weather, if the rain is accompanied by heavy winds, it is possible for specimens to be blown over, or certainly blown out of the vertical. It is almost certain that although the main stem of the Saguaro is slow growing, rarely more than 1 in. (2·5 cm) per year, lateral branches can grow more quickly in an effort to stabilise a specimen which has been moved by high winds.

The flowers are large up to 5 in. (12·5 cm) in diameter, quite waxy in appearance, and produced in May and June. Because of their scent they attract various insects and also the White-winged Dove and Longnose Bat, which, although in search of nectar, aid pollination by transferring pollen on their wings or body to flowers on another specimen. As a general rule bees are a common insect which help with pollination of daylight blooming plants, and moths for night blooming species; but doves or bats are far less common, but not unique.

The fruits which follow are large, and when they open out look almost like red flowers. Local Indians gather these fruits, and they make use not only of the pulp in the fruits for 'Cactus Jelly', but also of the seeds for the oil they contain. Various types of animal and bird life also come after the fruits, and in fact woodpeckers make holes within these Saguaros which when abandoned are taken over by the Elf Owl. The woodpeckers can sometimes cause the death of a giant Saguaro, if their boring activities are during wet weather, as disease can enter the cactus, and cause it to rot away within a year or so.

Plate 70 depicts four young Saguaros, three of which are growing beneath a Palo Verde tree. This beautiful yellow-flowered low-growing tree is a common feature in these parts, and it often plays host to young Saguaros for many years, giving them the necessary shade. As mentioned earlier, no cactus seedlings will survive long in the burning desert sun, but under a Palo Verde tree they stand a much better chance. However, this is not to the advantage of the Palo Verde tree eventually, as once the young Saguaros develop a large enough root-spread they absorb all the available moisture in that vicinity, which results in the death of the host. Up to perhaps 6 ft (1·8 m) in height, the Palo Verde tree will be alive, but beyond that stage its years are strictly numbered.

PLATES

67 and 68   General scenes in the Saguaro National Monument, taken in the western section, some 16 miles (26 km) west of Tucson.

69   Close-up of the top of an 8-ft (2·6-m) high specimen of *Carnegiea gigantea*.

70   Young specimens of *Carnegiea gigantea* growing beneath some desert trees, including the yellow-flowered Palo Verde tree.

71   A fine branched specimen of *Carnegiea gigantea*, where the main stem has become cristate. This was specially filmed near the San Xavier del Bac Mission, near Tucson. There is a flowering stem of *Fouquieria splendens* in the picture.

72   Here you can see the 'Gilded Flicker', one of two woodpeckers, which make their home within the Saguaro.

Location: Silver Bell Mountains in Pinal county some 40 miles (64·4 km) north-west of Tucson. To reach this spot on the map, we travelled by dirt road part of the way through quite flat desert scrub-land, where the Palo Verde tree was very prominent, also common Mesquite (*Larrea tridentata*) many covered with their yellow blossoms. Numerous young specimens of *Carnegiea gigantea* (Saguaro) were to be seen, also *Ferocactus wislizenii, Opuntia engelmannii*, and a few of the Cholla-type Opuntias. Higher up at around 3,500 ft (1,067 m) the slopes were quite steep and this is where we stopped.

These slopes, no matter how steep, were quite densely clothed with fine specimens of *Carnegiea gigantea* (Saguaro) but higher still they tended to thin out. The ground was very rocky indeed, and here we found again *Echinocactus horizonthalonius* (Blue Barrel or Eagle's Claw). Compared with the Texan specimens these were rather more globular and flattened, the spination being even stronger. Flowering plants of *Ferocactus acanthodes* (Compass or California Barrel) were scattered and varied in size. Some were only 10 in. (25 cm) across in bloom, while others such as the one in Plate 73 was 4 ft (1·3 m) high. Some of these large specimens grow quite precariously as their root system is not very large, but more often than not surrounding rocks help to prop them up. Higher up still, where it was quite inaccessible, some even larger specimens were visible. This species can attain a height of 10 ft (3·3 m) or so, and have a diameter of 18 in. (46 cm). The spines of these were much redder, compared to the more yellowish ones north-east of Apache Junction.

The common name Compass Barrel sounds intriguing. Its name is due to the fact that the top of the plant tends to point in a some-what southerly direction. This occurs with other cacti not only in this part of the world, but also with some of the other succulent plants native to the southern half of Africa. Those in the northern hemisphere which point to the south would point in a northerly direction if grown south of the equator.

Returning to our observations on this habitat, a very beautiful golden-yellow spined variety of *Echinocereus engelmannii* is to be found here which goes under the variety name *nicholii* (Nichol's Hedgehog). It is a clustering species which can form into very

E

large clusters of thirty or so stems, with single stems reaching 2 ft (60 cm) in height. The flowers are sometimes a little paler generally than *E. engelmannii*; the main species and the spines in some cases can be very pale yellow to almost white.

We again came across *Opuntia bigelovii* (Teddy Bear Cholla) on the south facing slopes. No old specimens were to be seen in this locality, just a fine stand of young ones which in time would become as large as those seen near Apache Junction.

PLATES

73 4-ft (1·3-m) high specimen of *Ferocactus acanthodes* in the Silver Bell Mountains, Pinal county, Arizona.

74 Close-up of the flowering top of *Ferocactus acanthodes*.

75 Close-up of *E. engelmannii* var. *nicholii* in flower.

76 *Echinocereus engelmannii* var. *nicholii* at the same locality. Highest stem about 15 in. (37·5 cm).

LOCATION J (PLATES: Nos. 77 to 87, MAP, p. 56)

Location: These illustrations were taken at three locations very near one another, just a few miles south and west of the San Xavier del Bac Mission near Tucson in Pima county, at about 2,500 ft (762 m). At both locations on level ground, among scattered specimens of *Cercidium floridum* and *microphyllum* (Palo Verde trees), the soil was of a very fine sandy nature. In fact, it was a very typical desert-scrub region, with the trees and bushes more prolific near a dry 'wash' area, as positions such as that would retain some moisture a little bit longer.

At the first of these locations, near Black Mountain, we came across some very fine examples of *Opuntia fulgida* var. *mamillata* (Smooth Chain Fruit Cholla) as scattered specimens about 5 ft (1·6 m) high, one of the Chollas with the chain fruit habit. You may wonder what this means, but if you examine Plate 77 of *O. fulgida* var. *mamillata* and Plate 78 of *O. fulgida* you will be able to see quite clearly a series of green fruits joined together. What has happened is that having flowered and produced a fruit, another flower bud has formed out of the unripe fruit. This can go on for many years, to such an extent that it is occasionally possible to find a chain of these unripe fruits up to a few feet (a metre or so) in length. Usually the fruits are sterile, but this is not always the case, and there are other species of Cholla-type Opuntias which do the same thing. It is possible to count up the number of years by means of these fruits, as normally only one flower is produced at one spot on the chain each year. If the fruits ripen, they will be eaten by birds who will then spread the seeds far and wide for germination at a later date. If they do not ripen, one or more of these unripe or sterile fruits will drop off, root and then grow into a new plant. It is in fact an additional means of survival for such species, as the normal spiny joints of these species are easily detached, and transported accidentally by animals, who later perhaps rub them off, when they take root and grow into further fine specimens. There were also numerous fine specimens of *Mammillaria microcarpa* (Fish-hook Pincushion), small globular or short-cylindrical cacti with prominent hook central spines. The top of one such specimen is shown on Plate 79. This species is very widespread, not only over a wide area of the southern half of Arizona, but also over the border into the adjoining state of California to the west, and southwards into the Sonoran part of Mexico.

Nearby, but usually more in the open, were two other low-growing cacti, *Coryphantha muehlenpfordtii* var. *robustispina* (Stout Needle 'Mulee') and *Echinocereus fendleri* var. *robustus* (Robust Hedgehog).

At our next location, in similar sandy-soil conditions but where the vegetation was somewhat thicker with various kinds of low-growing trees including the ever-present Palo Verde trees, were scattered specimens of *Ferocactus wislizenii* (Candy or Fish-hook Barrel). They varied from juvenile plants less than 10 in. (25 cm) to the occasional one not far short of 6 ft (2 m) in height, such as that shown in Plate 80. Specimens as high as 10 ft (3·3 m) have been found, and as these plants are far from being quick growing, one hesitates to hazard a guess at the great age a specimen like that would be. Two other illustrations depict the top of a 3-ft (91-cm) high specimen with fruits and a much smaller and younger specimen in bloom. Flowering occurs mainly in the summer and the fruits do not ripen until the winter. The specimen covered with ripe yellow fruits was filmed in May. We can only assume that in this particular area there were not many rodents or even deer, as otherwise they would have been eaten long before May came round.

At our last spot in this area, but in very similar surroundings, in addition to the very occasional *Carnegiea gigantea* (Saguaro), we found numerous fine specimens of *Opuntia fulgida* interspersed with some quite large Palo Verde trees. *O. fulgida* is more densely spined than the var. *mamillata* already referred to, giving it a much whiter appearance. Beneath these Opuntias ranging between 6 ft and 9 ft (2 m and 3 m), we found numerous clumps of another hooked spined *Mammillaria*—*M. thornberi* formerly perhaps better known as *M. fasciculata*. Two illustrations are included of it, one showing it in the shade of one of these Opuntias, and another as a close-up in flower. This species was only to be found in shady positions, never in full sun positions. From the many specimens examined, we noticed that often a section of a clump had died, yet that same area was being re-colonised among the dead remains of an earlier part of the same clump. Some of these clumps were quite large, 12 in.–18 in. (30 cm–46 cm) across, and could have more than a hundred heads, but these were in the minority.

Nearby either in the shade of the Palo Verde trees or in the open, we found what must be one of the most attractive of all the flat padded Opuntias, *O. santa-rita* (Purple Prickly Pear). It has gained this common name because of the wonderful colourful hues the pads, 6 in.–8 in. (15 cm–20 cm) in diameter, can attain. The pads are quite thin, and in fact do not store as much water as most of the other species are capable of doing. The pads are also almost spineless, except for a few long dark brown or even pink spines at the top. As the pads are so slim and almost spineless, thus having little protection, this species can not only suffer badly during long periods of drought, but in such hard times rodents will eat them. Cattle too will eat them when the need arises. Other species of Opuntia may be eaten by cattle, particularly if they are species with few spines, or their spines have been burned off. Some years ago a hybrid Opuntia was produced as cattle fodder with large fat succulent pads. This was spineless, but even so it is not the best replacement for more normal forms of grazing.

PLATES

77 *Opuntia fulgida* var. *mamillata* in habitat some 5 miles (8 km) south of Tucson, near Black Mountain.

78 *Opuntia fulgida* var. *mamillata*, showing a few 'chain fruits'.

79 A specimen of *Mammillaria microcarpa* about 3 in. (7·5 cm) across from beneath the earlier illustrated specimen of *Opuntia fulgida* var. *mamillata*.

80 5-ft 6-in. (1·7-m) high specimen of *Ferocactus wislizenii* being examined by a friend some 2 miles (3·2 km) south of the San Xavier del Bac Mission, near Tucson.

81 Close-up of the flowers of a young specimen of *Ferocactus wislizenii*.

82 The top of a 3-ft (91-cm) high specimen of *Ferocactus wislizenii*, with numerous ripe fruits.

83 The top of a 10-ft (3·3-m) high specimen of *Opuntia fulgida*, a few miles west of the San Xavier del Bac Mission, near Tucson. Note the well-developed chain fruit system.

84 Clumps of *Mammillaria thornberi* growing in the shade of *O. fulgida*.

85 Close-up of a clump of *Mammillaria thornberi* in bloom.

86 5-ft (1·6-cm) high clump of *Opuntia santa-rita* growing at the same locality as *O. fulgida* and *M. thornberi*.

87 Close-up of *O. santa-rita*.

LOCATION K (PLATES: Nos. 88 to 95, MAP, p. 56)

Location: North-east of Apache Junction on the border between Pinal county and Maricopa county and on the way to Canyon Lake, at an elevation of about 2,000 ft (600 m) in Arizona.

At this location it was possible to find nine different species of cacti, three of which grew on the lower sandy area. The commonest species was *Opuntia fulgida* (Jumping Cholla) with specimens varying from 12 ft (3·7 m) to 15 ft (5 m) in height. *Opuntia engelmannii* (Engelmann's Prickly Pear) consisted of bushes 4-ft (1·3-m) high. Compared with the Texan forms it had slightly longer spines, no doubt on account of the dryer locality. There were also a few scattered specimens of *Carnegiea gigantea* (Saguaro); these were mostly single unbranched stems below 20 ft (6·1 cm) in height. Compared with those in some areas these few specimens had quite thin stems.

On the lower slopes, on much rockier ground, the other six species were to be found, although the best specimens of *Ferocactus acanthodes* (Compass Barrel Cactus) were seen on the upper slopes, some growing out of sheer rock and measuring upwards of 6 ft (2 m) in height. These specimens were almost entirely pale pink to yellowish spined species, unlike the much redder spined species seen in the Silver Bells (see Plate 74).

The outstanding species on the lower slopes was *Opuntia bigelovii* (Teddy Bear Cholla), with its brilliant golden straw-coloured spines. As you can see by the illustrations (Plates 90 and 91) it is very densely spined, and the flowers are borne on the tips of the new joints. The old spines lower down on the plants become quite black. Most of the specimens here averaged between 6 ft and 8 ft (2 m and 2·6 m) in height. Interspersed between the dense colonies of *Opuntia bigelovii* were occasional specimens of *Opuntia acanthocarpa* var. *thornberi* (Thornber Cholla), most of them below 6 ft (2 m) in height, and having yellowish flowers.

Beneath these Opuntias, sometimes growing tightly around the main stem, were numerous specimens of a form of *Echinocereus engelmannii* (Strawberry Hedgehog Cactus) and another form of *Mammillaria microcarpa*. If you compare the one in Plate 92 with the other one of *Mammillaria microcarpa* (Plate 94), a marked difference can be noted. These two dwarf growing cacti are found beneath the Opuntias because the seeds had been passed in the faeces of a bird which had previously enjoyed the very delicious fruits of the Echinocereus and the Mammillaria; the young plants also enjoyed the shade given them by the 6-ft (2-m) high Opuntias.

PLATES
88 View showing single stemmed Saguaro in flower, backed by a mass of *Opuntia fulgida*.
89 View showing *Opuntia bigelovii* at bottom left. The best specimens of *Ferocactus acanthodes* were on the upper slopes.
90 A fine stand of *Opuntia bigelovii*.

91   Close-up of the top of one specimen of *Opuntia bigelovii* in bloom.

92   Close view shot, depicting a double headed *Echinocereus engelmannii* with one fruit, alongside which you can see three specimens of *Mammillaria microcarpa*.

93   Close-up of *Echinocereus engelmannii* in bloom.

94   Close-up of a three-headed specimen of *Mammillaria microcarpa*.

95   Close view of *Ferocactus acanthodes* (small specimen) in bloom.

LOCATION L (PLATES: Nos. 96 to 102, MAP, p. 56)

Location: Fish Creek Hill at an elevation of about 2,000 ft (600 m). This is on the Apache Trail not very far from Canyon Lake, where numerous specimens of *Carnegiea gigantea* (Saguaro) can be found growing down to the lakeside, such as the giant specimen some 40 ft (13·3 m) high seen on a promontory beside Roosevelt Lake, along with many other smaller cacti. The Apache Trail starts some 30 miles (48 km) east of Phoenix and goes through the Superstition mountains, through Fish Creek Canyon, and along the southern side of Canyon, Apache and Roosevelt Lakes, ending up at Globe.

Although this selected site is Fish Creek Hill, Plate 96 shows a very similar site near Canyon Lake. Note the sloping terrace formation which in parts is quite polished so that exploration has to be carried out carefully. Some of the ledges lower down the sides of these canyons are very narrow and great care has to be taken. The upper sloping terraces on Fish Creek Hill were very interesting because the plants that grew there were often only in shallow saucers of soil, and very stunted as with the specimens of *Fouquieria splendens* (Ocotillo). The occasional plant of a form of *Agave palmeri* (Blue Century Plant) were to be found together with numerous specimens of the narrow leaved *Agave toumeyana*. Beneath them one often found *Dudleya saxosa* var. *collomiae*, a small powder-blue-leaved succulent of dwarf habit related to the Echeverias. In addition specimens of *Echinocereus fendleri* var. *boyce-thompsonii* (Boyce-Thompson Hedgehog), a very variable and attractive member of the genus, as you can see in Plates 98 and 99. The

Dudleya and *Echinocereus* were also to be found in the open where they received full sun conditions, and given a few inches of soil both these species grew into large clusters. As you can see by the two illustrations of this *Echinocereus* a tremendous variation occurs in the spine colour and to a certain extent in the flower too. Specimens of this *Echinocereus* were found within 20 yards (18·3 m) of each other with spines ranging in colour from pure white to dark brown or nearly black in one case. This same species was also found nearer Globe on a hillside where there were numerous outcrops of white quartz. Here all the specimens of *E. fendleri* var. *boyce-thompsonii* were fairly uniformly pure white, and the plants of *Mammillaria microcarpa* (Fish-hook Pincushion) and a form of *Agave toumeyana* had much more white in their colouration than usual, so that all the plants blended into the hillside.

Returning to Fish Creek Hill, this *Echinocereus* was not to be found on the lower ledges, whereas the *Dudleya* was, even growing in dense shade among other shrubs. Usually in some shade, if not dense shade, another miniature succulent plant was to be found here, *Graptopetalum rusbyi*, a clustering species which is shown in close-up on Plate 101.

Another miniature we found here, but a true cactus, was a form of *Mammillaria wrightii* var. *wilcoxii*, often growing out of vertical cracks in the canyon wall, but usually in a position where it did not receive any shade. Many of the specimens were little over 1 in. (2·5 cm) in diameter. It is a densely spined variety as you can see by Plate 102. From six selected specimens, at one end of the line you had a plant that compared favourably with the var. *wilcoxii* and at the other end with var. *viridiflora*. The plant illustrated is more or less midway between these two varieties. As we have mentioned in Chapter 3 on the classification of cacti, it is difficult to know where you draw the line between one species or variety and another. In the case of *Mammillaria wrightii* (Wright's Pincushion), for instance, are they all forms of one species ranging in flower colour from magenta through to pink and green, especially bearing in mind that the spine count also varies from one plant to the next? It is only when you study a large number of specimens in the wild that this variation can be fully understood.

PLATES

96 Canyon near Canyon Lake on the Apache Trail, with flowering stems of *Fouquieria splendens* (Ocotillo) in the foreground.

97 A typical sloping terrace on Fish Creek Hill, with a plant of *Echinocereus fendleri* var. *boyce-thompsonii* at bottom left and a mixed patch of *Agave palmeri* and *Agave toumeyana* at top right.

98 Clustered specimen of *Echinocereus fendleri* var. *boyce-thompsonii*.

99 White-spined form of *Echinocereus fendleri* var. *boyce-thompsonii*.

100 Specimen of *Dudleya saxosa* var. *collomiae* Moran.

101 Close-up of flowering plant of *Graptopetalum rusbyi*.

102 Close-up of a form of *Mammillaria wrightii* var. *wilcoxii*.

LOCATION M (PLATES: Nos. 103 to 107, MAP, p. 56)

Location: About forty miles (64·4 km) south-west of Prescott, in Yavapai county, Arizona, not far from Yarnell and at an altitude of 4,700 ft (1,433 m). A rocky hillside, among dry grass about knee-high in parts, with huge granite boulders scattered everywhere. The situation was south-west facing, and there were a few scattered bushes or low growing trees.

Because of the higher altitude the specimens of *Opuntia basilaris* (Beaver Tail Prickly Pear) were in bloom, whereas specimens situated at around 3,000 ft (1,000 m) had finished blooming. This is a very attractive low-growing species, which to the average person looks quite harmless to touch compared with some of its relatives, which have very strong and obvious spines. *O. basilaris* has small clusters of brownish glochids, which are really minute spines, and these are barbed all the way up. Most Opuntias have glochids as well as the larger spines. *O. basilaris* tends to branch from the base of the previous pad, and this is why it remains very low-growing. In the heat of the summer the pads shrivel a lot, but they soon fill out again once the rains come. The specimens were growing among grass and some from crevices in the rock.

Another Opuntia to be found here was *O. chlorotica* (Clock-face Prickly Pear), a species which can reach up to 8 ft (2·6 cm) in height with a cylindrical main trunk. The pads of this species are almost circular, with golden glochids and spines, and it is certainly attractive. In this particular locality, few specimens exceeded 4 ft (1·3 m) in height, and the one illustrated (Plate 105) was growing out of a crevice in a huge boulder. This specimen was growing in partial shade, in fact the best ones were all in such a position but usually with their roots in the ground rather than in a crevice.

Among the dry grass two other low growing species were to be found, *Coryphantha vivipara* var. *arizonica* (Beehive Cactus) and *Echinocereus triglochidiatus* var. *melanacanthus* (White-spined Claret Cup). These two species are ones which can be found at much higher altitudes, up to 7,000 ft (2,133 m) or so—in fact this *Echinocereus* has been found at higher altitudes still, up to around 10,000 ft (3,047 m). These species grow on rocky hillsides in pine forest land, whereas at this location they are growing on dry grassland, not so many miles from desert areas, and at about their lowest altitude. Species such as these at higher altitudes would endure quite a few degrees of frost, and snow on occasions, but they cannot be termed as very hardy compared with such plants as *Neobesseya missouriensis* or *Echinocereus viridiflorus*.

Coryphantha vivipara var. *arizonica* can be solitary, but can branch into clusters that may measure 2 ft (60 cm) across, whilst *E. triglochidiatus* var. *melanacanthus* had been known to reach immense proportions, with as many as 400–500 heads per clump. It does not seem to be such a free blooming species as other members of the same genus, but the scarlet flowers stay open day and night for four to five days. One clump of this species was found growing high up on a huge granite boulder (Plate 103).

PLATES

103 *Echinocereus triglochidiatus* **var.** *melanacanthus* growing out of a crack in a 10-ft (3-m) high granite boulder, some 40 miles (64·4 km) south-west of Prescott, in Yavapai county, Arizona, near Yarnell at an altitude of 4,700 ft (1,433 m).

104 3-ft (91-cm) diameter clump of *E. triglochidiatus* var. *melanacanthus*.

105 *Opuntia chlorotica* growing from a crack in a granite boulder, and about 4 ft (1·3 m) high.

106   Fine flowering clump of *Opuntia basilaris*, which measured just over 2 ft (60 cm) in diameter.

107   Close-up of a 4-in. (10-cm) diameter clump of *Coryphantha vivipara* var. *arizonica*.

LOCATION N (PLATES: Nos. 108 to 113, MAP, p. 56)

Location: Near Alamo Crossing, some 25 miles (40·2 km) west of Yarnell, Yavapai county, at an altitude of about 3,000 ft (914 m).

This is a silvery sand desert area, where the most prominent species to be seen is *Yucca brevifolia* (Joshua Tree), some specimens of which are 25 ft (8·3 m) high. This is a slow-growing tree, which is better known further west in California, where there is the Joshua Tree National Monument. It is one of the Liliaceae.

In summer this is a hot desolate spot, and in almost pure sandy soil, plants really have to search for their moisture, so that these slow-growing Joshua trees must be a great age. Nearby *Opuntia basilaris* is to be found, but the specimens are very sad looking compared with those seen higher up near Yarnell.

Scattered specimens of *Opuntia kleiniae* var. *tetracantha* and *Opuntia acanthocarpa* were to be seen, and a low growing and densely spined form of *Echinocereus engelmannii* (Strawberry Hedgehog). Because of the poor soil and water scarcity, to say nothing of the heat in this rather desolate spot, it is to be expected that plants would be stunted, and so better equipped to combat the local conditions. The more densely spined a plant is, such as this *Echinocereus*, the better it can withstand excessive heat, as the spines cast a shadow over the body of the plant and help reduce transpiration. Generally these variations apply mainly to the vegetative structure of the plant, rather than to flower size, but with this particular form of *E. engelmannii* even the flowers are a little smaller, whether produced in habitat or in cultivation. This particular species has a wide distribution over four states of the USA as well as into the Sonora and Baja California parts of Mexico, so local variations are to be expected.

By diligent hunting beneath the Opuntias and Joshua Trees it is possible to find one of the large flowered, but quite small growing, species of Mammillaria—*M. tetrancistra* (Corky-seed Pincushion). This is a small columnar species, usually solitary and only to be found in areas where the rain-fall is small and in very well drained positions. In cultivation this is one of the more difficult species to cultivate, as great care has to be taken not to kill it with kindness! In the wild it does not seem to colonise as do many of the species of cacti. There are only scattered specimens over a wide area from Nevada and Utah through to Arizona and into California. This particular locality is the nearest to a true desert location of all the plants illustrated and described in this book.

PLATES

108 General view of this desert-like area near Alamo Crossing, some 25 miles (40·2 km) west of Yarnell, Yavapai county, at an altitude of about 3,000 ft (1,000 m). In addition to certain scattered specimens of various types of thorn bush, you can see one of the Cholla Opuntias (*O. acanthocarpa*) to the right of centre, a plant of *Echinocereus engelmannii* at bottom left, while standing out above the other bushes you can see a number of the smaller specimens of *Yucca brevifolia* (Joshua Tree).

109 Some 25-ft (8·3-m) high specimens of *Yucca brevifolia* (Joshua Tree).

110 Close-up of a flowering head of *Yucca brevifolia*.

111 A four-headed specimen of *Echinocereus engelmannii* with fruits on it, whilst lying nearby are the dried woody remains of sections of a Cholla Opuntia. These pieces are often collected by flower arranging enthusiasts and used in special desert flower arrangements, for which they are quite popular.

112 Close-up of a 3-in. (7·5-cm) high specimen of *Mammillaria tetrancistra*.

113 15-in. (37·5 cm) diameter clump of this form of *Echinocereus engelmannii* found near Alamo Crossing.

CALIFORNIA

LOCATION O (PLATES: Nos. 114 to 121, MAP, p. 74)

Location: Not far from the California/Mexico border in San Diego county, at a very low elevation and within fifty or so yards of the Pacific Ocean. This is a typical California Chaparral, except that the cacti only appear when the vegetation thins out; in fact one or two species are more prevalent on dry slopes and in rock crevices where there is practically no other vegetation at all.

This particular spot is very typical of many to be found in this coastal Chaparral region of California. It is only in the San Diego city area and southwards to the Mexican border and also down into the northern part of the Baja Californian peninsular part of Mexico, that *Bergerocactus emoryi* becomes a dominant feature wherever the great variety of low-growing shrubs and trees are thin on the ground. As a general rule, unless cacti have underground storage root systems from which they can regrow, they do not stand up to fires very well. Because of this most of the cacti are to be found where this general type of vegetation is more sparse, and where any fire would not be too intense or last sufficiently long to kill them.

Bergerocactus emoryi has quite slender stems, but is densely spined; it forms into sprawling and very dense bushes rarely even reaching 3 ft (91 cm) in height. The new growth of this species is a most attractive golden-yellow colour, but it pales with age and changes to a dirty colour on the old stem bases.

A number of other smaller and low-growing plants are often to be found near or even among some of these dense colonies of *Bergerocactus emoryi*. They include a short columnar hooked spined Mammillaria—*M. dioica*, which can sometimes be found as clusters, but single stems are more common, these rarely exceeding 6 in. (15 cm) in height. More than one species of *Dudleya* often grows in association with this *Mammillaria* in a location such as this. The *Dudleyas* are perennial succulent-leaved plants belonging to the Crassulaceae family, of which there are a great variety of species, particularly along the coastal region of southern California and also down into Baja California. Some species are solitary kinds, others low-growing free branching species, whilst the

SOUTHERN PART OF CALIFORNIA (showing county areas)

NEVADA

INYO

TULARE

KERN

SAN BERNARDINO

SANTA BARBARA

VENTURA

LOS ANGELES

PACIFIC OCEAN

San Bernardino

Los Angeles

RIVERSIDE

JOSHUA TREE NATIONAL MONUMENT

Palm Springs

Indio

SANTA CATALINA ISLAND

ORANGE

N

SAN DIEGO

Salton Sea

ARIZONA

SAN CLEMENTE ISLAND

Escondido

IMPERIAL

San Diego

El Centro

Tijuana

Mexicali

Yuma

BAJA CALIFORNIA IN MEXICO

leaf colour may be green or mealy-white in colour as in *Dudleya saxosa* var. *collomiae* from Arizona (Plate 100).

A smaller growing Barrel Cactus is also to be found here. Its habitat extends further north along the coast to about half the length of the coastline of San Diego county and inland to a limited degree. The name of this particular species is *Ferocactus viridescens* and, unlike its larger-growing relatives which we have described in the Arizona section, this species does not grow very tall. To see a specimen 12 in. (30 cm) high is far from common; more often than not it is a flattened globular species up to 15 in. (37·5 cm) or so in diameter. There are four prominent central spines, the lowest one of which is the most prominent. This species is quite variable, and in some cases these particular spines tend to be more flattened against the small radial spines and the body of the plant. Unlike the larger Barrel Cacti referred to earlier, specimens of this species can be seen in flower when only 4 in. (10 cm) or so in diameter. It tends to be found in slightly more open positions on slopes where the soil is of a gravelly or sandy type, obviously preferring a better drained position and more often than not in full sun.

A number of flat padded Opuntias are to be found along the southern Californian coastline and further inland, of which the various varieties of *O. littoralis* are the most common. Another species which has been confused for many years with *O. littoralis* is *O. oricola*. This species is found on the coast as far north as Santa Barbara county, but some fine stands were to be seen here in San Diego county as you can see by Plate 121. Under ideal conditions it can reach a height of 10 ft (3·3 m), but here few specimens exceeded 3 ft–4 ft (91 cm–1·3 m) particularly as in this illustration, where they grew on steeply sloping ground.

Opuntias, Bergerocacti and certain *Dudleyas* are also to be found on some of the islands nearby such as Santa Catalina and San Clemente.

PLATES

114 General habitat scene with *Bergerocactus emoryi* in the foreground, with typical California Chapparral vegetation behind.

115   Close view of a clump of *Bergerocactus emoryi*.
116   Flowering plant of a form of *Dudleya lanceolata*, with a 3-in.
       (7·5-cm) high plant of *Mammillaria dioica* nestling beneath
       it. To the left of this picture a few stems of *Bergerocactus
       emoryi* are again visible.
117   Close-up of a flowering stem of *Bergerocactus emoryi*.
118   A few fine specimens of *Ferocactus viridescens*, the largest
       being 10 in. (25 cm) in diameter.
119   One flowering specimen of *Ferocactus viridescens*, of the kind
       having rather more prominent central spines.
120   Close-up of the flowers of *Ferocactus viridescens* of the form
       where the spines are more flattened against the plant and
       are grey.
121   Steep hillside near the sea, with clumps of *Opuntia oricola*.

9.   OTHER DESCRIPTIONS OF PLATES

The remaining illustrations were filmed either in The Exotic
Collection, or in habitat, and include a number of other genera
not previously shown in the habitat sections. Plate 133 is included
to give readers an idea of what the desert can look like in the
spring, when there has been sufficient winter rain. The last six
illustrations in the book show some attractive lay-outs in cultivation, either outside where the climate is suitable, under glass or
as house plants.

PLATES

122   *Echinomastus johnsonii* in close-up, showing the very
           dense strong spination, typical of this genus. It is found
           mainly in western Arizona and eastern California, in
           rather dry locations where there is very good drainage,
           and this is also essential in cultivation.

123   Two illustrations of *Echinocereus fendleri* var. *bonkerae*,
and 124   from two locations not very far from Globe, Gila
           County, Arizona. These are two good examples of the
           variation that can occur: there are some differences

not only in spine length and colour, but the white form has much narrower petals.

125   *Echinocereus viridiflorus* is shown in flower in cultivation here, but the plants originated from New Mexico. It is a widespread species, some forms are much smaller-growing, and these are usually the ones which come from the colder regions.

126   Two illustrations of *Lophocereus schottii*, one depicting a
and 127   specimen in habitat about 15 ft (5 m) high, and the other showing a section of a flowering stem with a number of open flowers. Referring to the description of this plant on page 196, it is important to remember that the spines visible in the close-up illustration are quite different to those lower down the stems or on younger plants which have not reached flowering size. These other spines to which we refer are very short, not much over $\frac{1}{4}$ in. (0·6 cm) long and darker coloured. This plant grows in the USA at its northern-most habitat in Pima County, Arizona, within the protected confines of the Organ Pipe National Monument.

128   A single headed specimen of *Neobesseya similis*, about $1\frac{1}{2}$ in. (3·75 cm) in diameter, from Kansas. It is often solitary or sparingly branched, compared with the more southerly locations where growing conditions are better, then larger headed clumps are more common.

129   A very fine cluster of *Pediocactus knowltonii* from the Type Locality (original locality where it was dis-covered). This plant is about $2\frac{1}{2}$ in. (6·25 cm) across. It is found in Colorado and New Mexico.

130   Close-up of the top of a 3 in. (7·5 cm) diameter head of *Pediocactus simpsonii*, a species native to Kansas and many other colder states of the USA, where it grows in woodland settings sometimes up to 10,000 ft (3,300 m).

131   Close-up of the top of a 3-in. (7·5 cm) high plant of *Toumeya papyracantha*, often found in grassland, which is why it is so difficult to see, particularly when the

grass is dry. It comes from N.E. Arizona and adjoining New Mexico.

132 This shows the top of a $3\frac{1}{2}$ in. (8·75 cm) diameter specimen of *Utahia sileri*, one of the more difficult species to cultivate. It shares this distinction with *Coloradoa mesae-verdae*, as both need great care regarding drainage and watering to avoid rotting. It comes from Mohave County, Arizona and nearby Utah.

133 Although the only cacti visible in this picture are in the background, this shows how colourful the desert can be when some of the wildflowers, including native annuals come into bloom in the spring. A show like this only occurs in the spring, when there has been sufficient rain during the latter part of the winter to germinate the dormant seeds, which may have been waiting a year or so for the right conditions. Arizona Poppy (*Kalistroemia grandiflora*); Prickly Poppy (*Argemone platyceras*); Desert Marigold (*Baileya multiradiata*); Paper Flower (*Psilostrophe cooperi*); Ground Aster (*Townsendia arizonica*) are just a few of the wild flowers which help to colour the desert. This was filmed in Yuma county, Arizona, and the Pina Cata Mountains are over the border in Mexico (see map, page 56).

134 This shows a fine flowering specimen of *Nolina parryi* from California. *Nolinas* can be confused with *Dasylirions* and *Yuccas*, but they are easy to distinguish as *Nolinas* have long grass-like foliage. *N. parryi* is in fact the largest growing *Nolina* in the USA and the plant illustrated would measure over 9 ft (3 m) to the top of its flower spike.

135 This is just a small part of the garden of our friends Alan and Betty Blackburn in Tucson, Arizona, showing two superb cacti specimens growing together. The globular golden spined plants are *Echinocactus grusonii* from Mexico. Each head was about 16 in. (40 cm) in diameter. The common name of this plant is Golden Barrel Cactus. The clustered columnar cactus with

red flowers hails from Argentina and is known under two names, either *Trichocereus huascha* var. *rubra* or *Lobivia huascha* var. *rubra*.

136   This is another way of growing cacti, but a rather un- usual one, which is becoming increasingly popular in the USA, using large pieces of volcanic rock or lava rock. Having obtained a piece of rock, a few holes or cavities are made in it into which soil and young plants are planted. These can look most attractive in a Spanish-style patio garden. This same sort of idea can be done where volcanic rock is not available by using tufa rock, which has a similar texture to this black volcanic rock, but is creamy-white in colour and very light-weight.

137   A view in one of the five showhouses of The Exotic Collection, Worthing, England, where the raised bed method has been used for growing the cacti. These beds are only about 6 in. (15 cm) deep, and are made on 2 in. (5 cm) thick floor boarding, which has been covered with roofing felt (Ruberoid) to preserve the wood. The sides of the beds are made of rock-surfaced concrete blocks, with side drainage holes at intervals below them. These beds have been made at a comfort- able height for studying the plants. The distance from the foreground to the end of the greenhouse is 60 ft (20 m) which is just under half the length of this house. The width is 20 ft (6·6 m) and a height of just over 10 ft (3·3 m) at the north end, as it has been built on a slightly sloping site. The bed in the foreground con- tains only *Echinocerei* from the USA, while the beds in the background contain Mammillarias from Mexico, and many other genera from the same part of the world.

138   A view in our 120 ft (40 m) long lean-to greenhouse, where all the plants are in a pseudo-natural setting, with winding paths. These have been made to look slightly volcanic looking, using cement, and then spraying with iron-sulphate solution before it is dry to

give the colour effect. The plants visible here include various flat padded Opuntias; the narrow green leaves in the centre belong to *Agave filifera*, while the white striped ones belong to *Agave americana* var. *mediopicta* forma. *alba*, which was introduced into cultivation by us some twenty years ago.

139 An attractive bowl garden, with the following plants: On the left *Dudleya brittonii*, front centre *Echinocereus baileyi*, right *Hamatocactus setispinus*; the flat padded plant is a small form of *Opuntia rufida*. This arrangement is in a 8 in. (20 cm) diameter and 3 in. (7·5 cm) deep Dragon Pottery bowl from Wales, UK. The figure on the left is 9 in. (22 cm) high and is Capodimonte porcelain from Italy.

140 This is another bowl garden of ours, using the following plants. Front centre: *Astrophytum asterias*. Right: *Dudleya saxosa* var. *collomiae*. At the rear a clump of *Echinocereus melanocentrus*. The arrangement has been made using an oval 7½ in. (18 cm) long and 2 in. (5 cm) deep pottery bowl by Moira Follery Ltd, England.

10. NATIONAL PARKS AND MONUMENTS WHERE CACTI CAN BE SEEN

As we have said in Chapter 5, the need for conservation is urgent; otherwise in a matter of a few years many species of plant-life will have disappeared for all time, and future generations will only know of them by means of pickled or dried herbarium specimens in some museum. Fortunately there are in the USA a large number of places where plants (and animals too) are still to be seen in their natural environment, in various National Parks and Monuments. In addition, there are also botanical gardens and other similar places, such as the cactus garden attached to the Judge Roy Bean Visitor Center in Langtry, Val Verde county, Texas. Here a small but attractively laid out cactus garden was started in 1968 by the Texas Highway Department, making a valuable addition to this unusual museum.

We are giving information on eight places, each falling into one of these categories, along with details on how you can obtain additional assistance and information by writing to the super-intendents at the addresses given. We have not included other botanical gardens in the USA—to do so would fill a book in itself! —but only those where part of the garden is virgin desert or desert-scrub. The Judge Roy Bean Visitor Center and cactus garden is included because it is unusual. It is a wonderful example to museums in many parts of the world where similar gardens could be added which would introduce visitors to some of the native flora of that particular area.

We are writing about these places in the same order as we dealt with the habitat locations of cacti from east to west, that is from Texas, New Mexico, Arizona through to California.

JUDGE ROY BEAN VISITOR CENTER

The primary object of the Texas Highway Department was to preserve a piece of the past—The Jersey Lilly Saloon where less than a hundred years ago Judge Roy Bean represented the Law west of the Pecos. However, in addition to preserving the old saloon, and building the Visitor Center, where visitors can learn about and relive the tough days of the past, a very attractive garden has been laid out adjoining it.

The Judge Roy Bean Visitor Center is situated in the very small town of Langtry in Val Verde county, Texas, only a mile or so from the Mexican border. It is reached off Highway 90, some 50 miles (80 km) north west of the Amistad Dam and Recreation Area.

This small but attractively laid out garden has been made using solely plants of the region. As well as a range of different *Opuntias* (Prickly Pear and Cholla types) and smaller Texan species such as *Epithelanthas*, *Neolloydias*, etc. there are some very fine specimen Yuccas, and some beautiful clumps of *Hesperaloe parviflora*, often called the red-flowered Yucca. Spring is really the best time for a visit, as many species will be in bloom, including some of the Yuccas. We saw the garden when it was quite newly opened. Within a few years it will be well established and will, we hope

encourage many people to grow some of these same plants themselves. Even in parts of the world where winter temperatures can
fall to twenty degrees below freezing or more on the Fahrenheit
scale, a number of species of Yuccas can grow, also the Hesperaloe.
When we were in Langtry the temperature was over 100°F (38°C)
showing how adaptable these plants are, provided you have the
right species.

FURTHER INFORMATION: The Superintendent, Judge Roy Bean
Visitor Center, Langtry, Texas, USA.
Texas Highway Department, Travel and
Information Division, Austin, Texas,
78703, USA.

BIG BEND NATIONAL PARK

Part of the photography of the Texan section of this book was
taken elsewhere in Brewster county, this is also the county in
which the Big Bend National Park is situated, with its southern
part flanked by the Rio Grande river on the border with Mexico.
On this border with Mexico the Rio Grande has cut canyons
which are 1,500 ft (500 m) deep. Two canyons which can be
reached without too much difficulty are Santa Elena to the west
and Boquillas to the south-east.

Three paved roads approach the Big Bend National Park; the
likely ones to be used are No. 118 south from Alpine, via Study
Butte just outside the Park, and No. 385 south from Marathon.

This park covers a very wide area as you can see from the map
on page 41 and it offers you the chance, at the right time of year,
of seeing the desert in all its glory, as over 1,000 different plants
grow there. Some 60 or so cacti are included in this figure, and
depending on the species flowers can be expected from the middle
of February through to May. Certain species flower even later
than that, for instance *Ariocarpus fissuratus* (Living Rock Cactus)
which blooms in the autumn (fall). Some of the other desert
plants also grow here; this park is within the Chihuahua Desert,
so *Agaves* (Century Plant) send forth their telegraph-pole sized
flower spikes from July onwards, while earlier in the year a

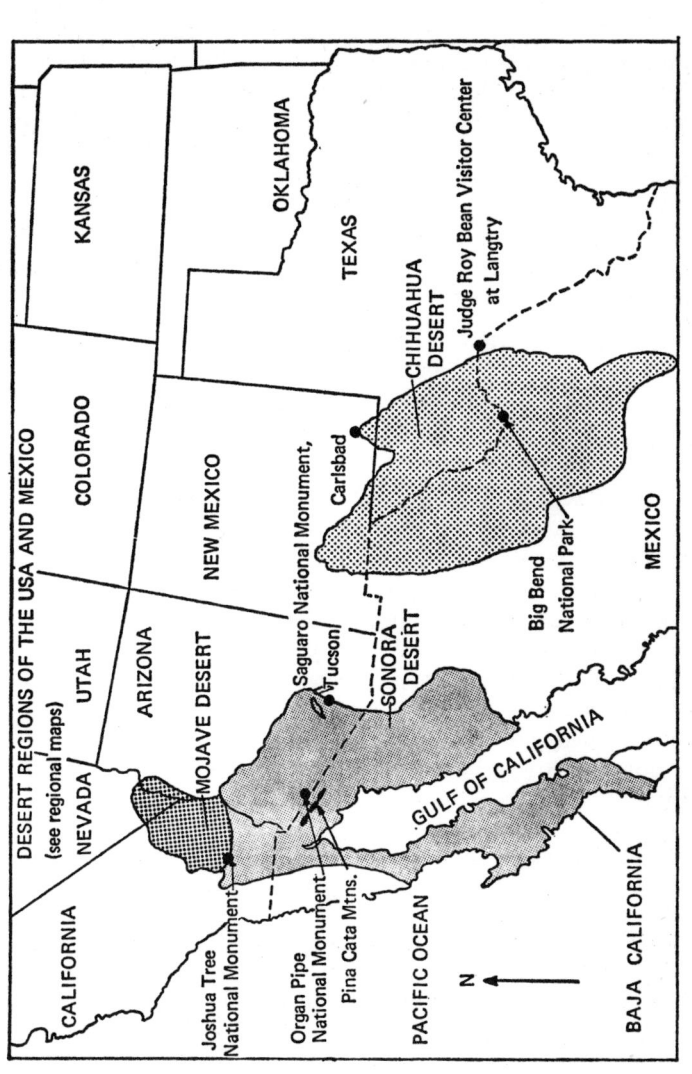

DESERT REGIONS OF THE USA AND MEXICO
(see regional maps)

number of species of Yuccas can be seen in bloom. There are also a great variety of wild flowers, some of them annuals, which can make a visit one to be remembered. Many of these bloom very early in the year, long before the heat of summer arrives, when temperatures can soar beyond the 100°F (38°C) mark.

If you contact the Park Headquarters when you arrive, or in advance, certain useful handouts will be supplied. The personal assistance and advice also available will help to make your visit a memorable and interesting one.

FURTHER INFORMATION: The Superintendent, Big Bend National Park,
Panther Junction,
Brewster County, Texas, USA.

CARLSBAD CAVERNS AND GUADALUPE MOUNTAINS NATIONAL PARKS

Despite the fact that the Carlsbad Caverns are in New Mexico in the south-east corner, and the Guadalupe Mountains National Park is in Texas, with a common boundary east to west along the New Mexico/Texas border, we are treating them as one.

The Carlsbad Caverns have been very well known for a long time and need little introduction as regards the fantastic subterranean geological formations, but above ground it is possible to see quite a variety of desert plants growing in their wild state within the confines of the park. In addition a Desert Natural Trail has been constructed, which with the aid of a booklet can be followed very easily for about a mile, when visitors are introduced to a fascinating variety of desert plants along with their common and botanical names.

This park and that of the Guadalupe Mountains are situated almost at the northern-most extremity of the Chihuahuan desert, where the flora is very similar to that seen further south in the Big Bend country. Ocotillo, Yuccas and Prickly Pear abound, along with a variety of the Hedgehog Cacti (*Echinocereus*) and other smaller-growing species belonging to such genera as *Mammillaria, Coryphantha, Escobaria sneedi* and *leei*; also *Epithelantha micromeris*

(Plate 41). Guadalupe Mountain itself is 8,751 ft (2,748·2 m) high, and there are other peaks only a little lower, so that a range of vegetation zones can be seen.

The Guadalupe Mountains were formed some 250 million years ago. Originally an area of some 10,000 square miles extending from north of Carlsbad right down to the Glass Mountains near Alpine in the Big Bend area was a vast inland sea. Eventually, through extensive movements of the earth's crust, the Guadalupe Mountains appeared, and then, through wind and rain, the softer parts eroded away leaving spectacular mountain and canyon scenery, with cliffs 2,000 ft (600 m) high. McKittrick Canyon is one such spot to be visited.

FURTHER INFORMATION: The Superintendent,
Carlsbad Caverns and Guadalupe
Mountains National Parks,
3225, El Paso Road,
Carlsbad, New Mexico 88220, USA.

CARLSBAD ZOOLOGICAL AND BOTANICAL STATE PARK

This was only opened in 1971, but is fast becoming one of the finest of its kind, devoted not only to animal life, but also to plant life, with special reference to cacti and succulent plants. It is situated on the north-western edge of the City of Carlsbad, New Mexico, in the Ocotillo Hills.

In addition to a fine display of cacti growing in the open-air, and possessing species from other parts of the world as well as those from the south-western part of the USA, there is also an under-cover display in the Succulent Cactus Pavilion, situated near the Duck Pond, containing those species which need a little more protection, even though Carlsbad has a relatively mild winter climate.

A place such as this can interest all members of the family, no matter to what group they belong, as unusual plants can be seen as well as a great variety of animal and bird-life, almost alongside each other.

FURTHER INFORMATION: The Superintendent,
Carlsbad Zoological and Botanical State
Park,
Carlsbad, New Mexico 88220, USA.

SAGUARO NATIONAL MONUMENT

You will already know much about *Carnegiea gigantea* (Saguaro) from illustrations and other details elsewhere in this book, and know that this giant cactus can be seen in places other than Saguaro National Monument. However, a visit to the Monument is a 'must' for anyone, because in addition to seeing the *Carnegias* and the other cacti which abound here, the exhibits at the Visitor Center are of outstanding interest. These have been produced in a very clear manner, so that they can easily be understood and appreciated. The various displays show not only how the land was formed, and how the various inhabitants (flora and fauna) have adapted to it, but also the life-cycle of the Saguaro Cactus. It shows how the Papago Indians have harvested its fruits for generations, and that within the stems of this giant cactus some kinds of birds live, such as the 'Gilded Flicker' and the 'Gila Woodpecker'. There are graded dirt roads in the western section and an $8\frac{1}{2}$-mile (13-km) macadam-surfaced loop road 'Cactus Forest Drive' in the eastern section. The Rincon Mountains contain about 65 miles (104 km) of clearly marked hiking trails, along with 13 miles (21 km) of trails in the Tucson Mountain section.

In the Rincon Mountains you can hike to an elevation of 8,666 ft (2,804·6 m) on Mica Peak, and in the process climb about 6,000 ft (2,000 m) through six distinct biotic communities from desert-scrub with the saguaros to the fir forest on the northern slopes of the Rincons. Camping permits must be obtained for overnight hikes.

FURTHER INFORMATION: The Superintendent,
Saguaro National Monument,
P.O. Box 17210, Tucson, Arizona
85710, USA.

DESERT BOTANICAL GARDEN IN
PAPAGO PARK

This 150-acre desert garden is situated off E. Van Buren St in Phoenix, Arizona, and is devoted entirely to desert plants from the many arid regions of the world. So in addition to seeing some species of cacti just as you find them in the 'wild', you can also see the great range of species to be found not only in North, Central and South America, but also in Africa, Arabia, Madagascar and elsewhere.

An Annual Cactus Show is staged in late February each year, with plant competitions, photographic and painting competitions; during the rest of the year lectures are given at regular intervals.

FURTHER INFORMATION: The Superintendent,
Desert Botanical Garden,
5800, E. Van Buren St.,
Box 5415 Papago Park,
Phoenix, Arizona 85010, USA.

ARIZONA–SONORA DESERT MUSEUM

This is situated in Tucson Mountain Park, just to the south of Tucson Mountain Section of the Saguaro National Monument. In addition to seeing a wonderful array of animals and birds, from mountain lions, rattlesnakes and scorpions to native birds or vampire bats, there are well-marked trails along which you can walk. Here you will find a display of native cacti, all clearly labelled. The feature of the Arizona–Sonora Desert Museum is that no matter whether the live exhibits are animals or plants, they all come from this area of the Sonora Desert only. In addition there are also some very interesting and educational geological displays.

FURTHER INFORMATION: The Superintendent,
Arizona–Sonora Desert Museum,
Tucson Mountain Park,
Tucson, Arizona, USA.

ORGAN PIPE CACTUS NATIONAL MONUMENT

Although *Lemaireocereus thurberi* (Organ Pipe) is to be found outside the boundaries of the Monument, there is one cactus species which is only found in the USA within this area, and this is *Lophocereus schottii* (Senita or Old One). It is a species which grows here along with *Lemaireocereus thurberi*, at their northern-most known locality, being far more common on the Sonoran and Baja California regions of Mexico. So, even if for no other reason, a visit here is worthwhile, but actually within this 500-square-mile (12,950-hectares) piece of the Sonoran Desert, you will find many other plants of interest, and wonderful scenery.

The Organ Pipe National Monument is located in Pima county, Arizona, on the border with Mexico. The usual method of getting there is either from the north on Highway 85 from Gila Bend, via Ajo and turning down to the Monument at a town called Why, or from the east on Highway 86 from Tucson, via Robles Juncton, Sells, through to Why. On the southern-most extremity of the Monument at Lukeville, one can cross over into Mexico via Sonoita.

It is possible to take a 21-mile (34-km) drive through the Ajo Mountains within the Monument and see a fine range of cacti including many species already illustrated in this book such as *Opuntia bigelovii* (Teddybear Cholla), *Echinocereus engelmannii* var. *nicholii* (Nichol's Hedgehog) and, if you are lucky, *Mammillaria tetrancistra* (Plate 112). Alternatively, a longer drive through the Puerto Blanca Mountains can be made, and then, with only the minimum of walking, fine stands of *Lophocereus schottii* (Senita or Old One) can be seen. However, if you plan on hiking off the beaten track, do take the advice of the Superintendent and his staff, who will help you to plan it, so that it can be done safely. Temperatures climb well over 100°F (38°C) in summer, so even within the confines of the Monument, you could run into trouble without prior help.

You can obtain either when you get there, or in advance from the Monument's headquarters, various handouts, which include details of the species of all forms of plant-life to be found there, and similar information on the types of animal and bird-life that live there.

FURTHER INFORMATION: The Superintendent,
Organ Pipe Cactus National Monument,
Box 38, Ajo, Arizona 85321, USA.

JOSHUA TREE NATIONAL MONUMENT

We have already mentioned earlier that *Yucca brevifolia* (Joshua Tree) is to be found on the western extremities of Arizona, but it is also found in Nevada, and in the Riverside and San Bernardino counties in California. The Colorado River is in fact the dividing line between California and Arizona, so that these unusual members of the family Liliaceae are to be found on both sides of it.

The Joshua Tree National Monument was initiated by President Roosevelt in 1936, with 825,340 acres (334,287 hectares) set aside, but if it had not been for the earlier work of Mrs. A. Sherman Hoyt, Founder and President of the Desert Conservation League, a few years more might have elapsed before this monument or park was set aside.

The scenery within the park is very dramatic. Some areas consist of giant boulders, many perched precariously on top of each other, as though some 'giant' had a field day a long time ago. In addition to the Joshua Trees which often line the road, you will also find desert willows in certain areas. There are other areas where various species of the Cholla-type Opuntias are to be seen along with the usual mesquite and creosote bushes, which are commonly and erroneously referred to as 'greasewood'. The Joshua Tree National Monument is so situated that the northern section lies within the Mojave Desert, whereas the south-eastern section is within the Colorado desert. You will remember from earlier on that the Colorado desert is really no more than the northern extension of the Sonora desert which reaches up from Mexico. Two trees which appear here are *Dalea spinosa* (Smoke Tree) and *Washingtonia filifera* (California Fan Palm).

If you make a tour of the Joshua Tree National Monument, in addition to seeing these weird trees, you will also see quite a range of other kinds of plant life.

A forty-mile (64·4-km) tour will enable you to do this, but no visit is really complete without going to Salton View at just over

5,000 ft (1,600 m), from where you can not only take in the wonderful sight of these peaks (San Jacinto at over 10,000 ft (3,047 m) and San Gorgonia at over 11,000 ft (3,350 m), but also the view across the valley floor as far as the Salton Sea, in addition to these peaks, which can be snow-covered, as can much of the monument in winter.

The Joshua Tree National Monument can be reached from the north, through the town of Twentynine Palms where the Monument headquarters is to be found, or from the south off Highway 10. The Joshua Trees and other species of Yuccas to be found in the Monument usually bloom, depending on the winter rainfall, between late March and on into May, whilst the cacti start in March and bloom into August.

FURTHER INFORMATION: The Superintendent,
Joshua Tree National Monument,
P.O. Box 875,
Twentynine Palms,
California 92277, USA.

NOTE: We would ask you when visiting one of these places, or any other similar ones, not to touch plants, and where cars are allowed, only to use them where permitted. There is nothing worse than seeing car tracks across virgin desert, as they may well remain for a long time.

Needless to say, do not leave litter around or light fires, except in the right places. Many of these parks and monuments will supply campers with full details of where fires can be lighted, but usually all fuel must be taken with you in the same way as you have to take the water along too. The addresses have been given to help you enjoy these places, so please make use of them.

We have deliberately avoided mentioning entrance fees, where this applies, as they can change, but where charges are made the money will go to help maintain these monuments and parks and so conserve at least a small part of nature for future generations.

THE COLOUR PLATES

1   View across the Mesquite north-west of Rio Grande City, Texas

2   *Homalocephala texensis* in habitat ($\times \frac{1}{6}$)

3   Close-up of the flowers of *Homalocephala texensis* (× ⅔)

4   *Escobaria runyonii* (× 1½)

5 *Ancistrocactus scheerii* ($\times$ 1$\frac{1}{2}$)

6 *Dolichothele sphaerica* ($\times 1\frac{1}{3}$)

7 *Thelocactus bicolor* ($\times 1\frac{1}{3}$)

8   *Astrophytum asterias* in habitat ($\times\frac{1}{5}$)

9   Three specimens of *Lophophora williamsii* in habitat ($\times\frac{1}{5}$)

10 *Astrophytum asterias* in flower (×2)

11 *Lophophora williamsii* (×4)

12 *Opuntia schottii* (×1½)

13 *Wilcoxia poselgeri* (×2)

14 *Yucca thompsoniana* in the Mesa country, Texas

15 *Coryphantha echinus* in habitat

16 *Coryphantha echinus. Right:* juvenile form, once known as *C. pectinata* ($\times \frac{1}{3}$)

17 *Dasylirion leiophyllum* south-east of Marathon, Texas

18   *Echinocereus triglochidiatus* var. *octacanthus* ($\times\frac{1}{10}$)

19   *Echinocereus triglochidiatus* var. *octacanthus* ($\times 1$)

20 *Hamatocactus uncinatus* growing among bushes of *Koeberlinia spinosa* ($\times \frac{1}{8}$)

21 *Hamatocactus uncinatus* ($\times$ 1)

22 *Hamatocactus uncinatus*, very long central spined form ($\times \frac{3}{4}$)

23 *Neolloydia conoidea* growing in bush of *Koeberlinia spinosa*

24 Fine clustered specimen of *Neolloydia conoidea*

25   *Neolloydia conoidea* (× 2½)

26 View 60 miles (96 km) south of Alpine, Brewster County, Texas

27 *Left: Echinocereus stramineus* in bloom, *bottom right: Escobaria tuberculosa*, and *top right: Opur leptocaulis* ($\times \frac{1}{25}$)

28   B. M. Lamb measuring a specimen of *Echinocereus stramineus* at the same locality

29   *Echinocereus stramineus* ($\times\frac{1}{10}$)

30   Fruit of *Echinocereus stramineus* ($\times 1\frac{1}{2}$)

31   *Opuntia engelmannii* $(\times\frac{1}{25})$

32   Orange-flowered form of *Opuntia engelmannii* $(\times\frac{1}{10})$

33   *Opuntia schottii* var. *grahamii* (×$\frac{1}{12}$)

34   *Echinocactus horizonthalonius* (×$\frac{1}{10}$)

35 *Yucca torreyi* south of Alpine, Texas

36 *Echinocactus horizonthalonius* ($\times$ 1)

37 *Ariocarpus fissuratus* at the same location as *Y. torreyi* ($\times \frac{1}{8}$)

38 *Ariocarpus fissuratus* (× 1½)

39 *Epithelantha bokei* in flower, with dormant specimen of *Selaginella lepidophylla* (× 1)

40 *Epithelantha bokei* (×4)

41 *Epithelantha micromeris* (×3)

42   *Mammillaria pottsii* $\times 2\frac{1}{2}$)

43   *Fouquieria splendens* in the Cuesto del Burro mountains

44   *Fouquieria splendens* ($\times \frac{1}{3}$)

45 *Opuntia imbricata* on the southern slopes of the Cuesto del Burro mountains ($\times\frac{1}{30}$)

46 *Echinocereus dasyacanthus* growing with *Agave lophantha* var. *poselgeri* ($\times\frac{1}{6}$)

47 *Echinocereus dasyacanthus*, a very old specimen ($\times \frac{2}{3}$)

48 *Hamatocactus hamatacanthus* surrounded by *Euphorbia antisyphilitica* ($\times\frac{1}{12}$)

49 View of the Rio Grande near Candellaria

50 *Peniocereus greggii* ($\times \frac{1}{15}$)

51 Flowers of *Peniocereus greggii* ($\times \frac{1}{2}$)

52   Stem and fruit of *Peniocereus greggii* ($\times 1\frac{1}{4}$)

53   *Echinocereus dubius (right)* and Mesquite bush *(left)* where a specimen of *Peniocereus greggii* was found ($\times \frac{1}{20}$)

54   *Echinocereus dubius* ($\times$ 1)

55   View in Pima county, Arizona, showing *Ferocactus wislizenii (left)* and *Fouquieria splendens*

56 Closer view, showing specimens of *Echinocereus rigidissimus*, including one in flower

57 *Echinocereus rigidissimus* in habitat ($\times\frac{1}{5}$)

58   *Echinocereus rigidissimus*, close-up of flowers (× ⅔)

59   Red-spined form of *E. rigidissimus* with one flower bud forming (× 2)

60   *Mammillaria heyderi* var. *macdougalii* ($\times \frac{1}{6}$)

61   Close up of *M. heyderi* var. *macdougalii* ($\times \frac{1}{2}$)

62 *Lemaireocereus thurberi* in south-west Arizona

63 Flower of *Lemaireocereus thurberi* ($\times \frac{1}{2}$)

64   *Lemaireocereus thurberi* ($\times\frac{1}{15}$)

65 *Ferocactus covillei* ($\times \frac{1}{12}$)

66 Flowers of *Ferocactus covillei* ($\times \frac{1}{3}$)

67 and 68 General views of the Saguaro National Monument, taken west of Tucson

69 *Carnegiea gigantea* ($\times \frac{1}{8}$)

70 Young specimens of *Carnegiea gigantea* beneath Palo Verde trees ($\times \frac{1}{40}$)

71   Cristate specimen of *Carnegiea gigantea* ($\times \frac{1}{15}$)

72   The 'Gilded Flicker', a woodpecker that lives in the Saguaro ($\times \frac{1}{3}$)

73 *Ferocactus acanthodes* seen in the Silver Bell Mountains, Arizona ($\times\frac{1}{25}$)

74 *Ferocactus acanthodes* ($\times\frac{1}{2}$)

75 *Echinocereus engelmannii* var. *nicholii* in flower (× 1½)

76   *E. engelmannii* var. *nicholii* ($\times \frac{1}{6}$)

77   *Opuntia fulgida* var. *mamillata* near Black Mountain ($\times \frac{1}{30}$)

78 *Opuntia fulgida* var. *mamillata* ($\times \frac{1}{6}$)

79 *Mammillaria microcarpa* ($\times 1\frac{1}{2}$)

80   *Ferocactus wislizenii* near Tucson

81   *Ferocactus wislizenii* in flower ($\times \frac{3}{4}$)

82   *Ferocactus wislizenii*, showing ripe fruits ($\times \frac{1}{10}$)

83   *Opuntia fulgida*. Note the well-developed chain fruit system ($\times \frac{1}{6}$)

84 *Mammillaria thornberi* growing in the shade of *O. fulgida* ($\times\frac{1}{10}$)

85 *Mammillaria thornberi* ($\times 2$)

86   Clump of *Opuntia santa-rita* ($\times\frac{1}{16}$)

87   *Opuntia santa-rita* ($\times\frac{1}{5}$)

88 Saguaro with mass of *Opuntia fulgida*

89 View showing *Opuntia bigelovii (bottom left)*

90 *Opuntia bigelovii* ($\times \frac{1}{30}$)

91 Top of *Opuntia bigelovii* in bloom

92 Double-headed *Echinocereus engelmannii* with one fruit, and three specimens of *Mammillaria microcarpa* ($\times \frac{1}{8}$)

93 *Echinocereus engelmannii* ($\times \frac{1}{4}$)

94 *Mammillaria microcarpa* ($\times$ 1)

95 *Ferocactus acanthodes* ($\times\frac{1}{2}$)

96 *Fouquieria splendens* (Ocotillo) near Canyon Lake on the Apache Trail

97 *Echinocereus fendleri* var. *boyce-thompsonii* (bottom left), and *Agave palmeri* and *Agave toumeyana* (top right) growing at Fish Creek Hill

98 *Echinocereus fendleri* var. *boyce-thompsonii* ($\times \frac{1}{4}$)

99 White-spined form of *E. fendleri* var. *boyce-thompsonii* ($\times \frac{3}{4}$)

100　*Dudleya saxosa ssp. collomiae* ($\times \frac{1}{4}$)

101　*Graptopetalum rusbyi* ($\times$ 1)

102 *Mammillaria wrightii* var. *wilcoxii* ($\times 2$)

103 *Echinocereus triglochidiatus* var. *melanacanthus* growing out of a crack in a granite boulder ($\times \frac{1}{20}$)

104 *Echinocereus triglochidiatus* var. *melanacanthus* ($\times \frac{1}{36}$)

105 *Opuntia chlorotica* ($\times \frac{1}{25}$)

106   *Opuntia basilaris* ($\times \frac{1}{6}$)

107   *Coryphantha vivipara* var. *arizonica* ($\times 1$)

108 View of the area near Alamo Crossing

109 A fine stand of *Yucca brevifolia* (Joshua Trees) ($\times\frac{1}{60}$)

110   *Yucca brevifolia* ($\times\frac{1}{4}$)

111   *Echinocereus engelmannii* and the woody remains of a Cholla Opuntia ($\times\frac{1}{6}$)

112 *Mammillaria tetrancistra* (×2)

113   *Echinocereus engelmannii* ($\times \frac{1}{3}$)

114   California Chapparral vegetation with *Bergerocactus emoryi* in the foreground

115 *Bergerocactus emoryi* ($\times\frac{1}{20}$)

116 *Dudleya lanceolata, Mammillaria dioica* and *Bergerocactus emoryi* ($\times\frac{1}{10}$)

117 *Bergerocactus emoryi* in flower ($\times\frac{1}{2}$)

118 *Ferocactus viridescens* ($\times \frac{1}{10}$)

119 *Ferocactus viridescens* ($\times \frac{1}{2}$)

120   Flowers of *Ferocactus viridescens* ($\times 1\frac{1}{2}$)

121   Steep hillside near the sea with clumps of *Opuntia oricola*

122 *Echinomastus johnsonii* (× 1)

123 *Echinocereus fendleri* var. *bonkerae* (× ⅓)

124   *Echinocereus fendleri* var. *bonkerae* ($\times \frac{1}{3}$)

125   *Echinocereus viridiflorus* ($\times \frac{2}{3}$)

126 *Lophocereus schottii* ($\times \frac{1}{40}$)

127   *Lophocereus schottii*, part of flowering stem (× ⅔)

128 *Neobesseya similis* (×1)

129 *Pediocactus knowltonii* (×2)

130   *Pediocactus simpsonii* (× 1½)

131   *Toumeya papyracantha* (× 3)

132 *Utahia sileri* (× 1)

133 The desert in springtime. Yuma County, Arizona

134 *Nolina parryi* ($\times\frac{1}{15}$)

135 *Echinocactus grusonii* (Golden Barrel Cactus) and *Lobivia (Trichocereus) huascha* var. *rubr* ($\times\frac{1}{20}$)

136 Display of cacti planted in cavities made in a volcanic boulder ($\times\frac{1}{10}$)

137   Opuntias and Agaves in the landscaped greenhouse ($\times \frac{1}{20}$)

138   The raised bed method of growing cacti, seen at the Exotic Collection ($\times \frac{1}{40}$)

139 Bowl garden with *Dudleya brittonii* (left), *Echinocereus baileyi* (centre), *Hamatocactus setispinus* (right) and *Opuntia rufida* ($\times \frac{1}{3}$)

140 This bowl garden contains *Astrophytum asterias* (front), *Dudleya saxosa* var. *collomiae* (right) and *Echinocereus melanocentrus* ($\times \frac{1}{3}$)

11. SIMPLIFIED CLASSIFICATION OF GENERA

following the system used by the Authors in their series *The Illustrated Reference on Cacti and Other Succulents* (Blandford).

Family. CACTACEAE
kăktā′sĭē

Group	Genera	Phonetic pronunciation
OPUNTIA	Opuntia	ŏpōō′ntĭă
CEREUS	Acanthocereus*	ăkănthōsē′rĭŭs
	Bergerocactus	bĕr′jĕrōkă′ktŭs
	Carnegiea	kărnē′gĭă
	Echinocereus	ĕ′kĭnōsē′rĭŭs
	Lemaireocereus	lĕmā′rĭōsē′rĭŭs
	Neoevansia	nēōvă′nsĭă
	Peniocereus	pĕ′nĭōsē′rĭŭs
	Wilcoxia	wĭlkŏ′ksĭă
PILOCEREUS	Lophocereus	lŏ′fōsē′rĭŭs
ECHINOCACTUS	Ancistrocactus	ănsĭstrōkă′ktŭs
	Ariocarpus	ărĭōkā′rpŭs
	Astrophytum	ăstrōfē′tŭm
	Coloradoa	kŏ′lŏrădō′ă
	Echinocactus	ĕ′kĭnōkăktŭs
	Echinomastus	ĕ′kĭnōmă′stŭs
	Ferocactus	fĕ′rōkă′ktŭs
	Hamatocactus	hă′mătōkă′ktŭs
	Homalocephala	hō′mălōsĕ′fălă
	Lophophora	lŏ′fōfo′ră

	Pediocactus	pē'dĭōkă'ktŭs
	Sclerocactus	sklĕ'rōkă'ktŭs
	Thelocactus	thĕ'lōkăk'tŭs
	Toumeya	tōo'mă'yă
	Utahia	yōota'hĭă

MAMMILLARIA	Coryphantha	kŏrĭfăn'thă
	Dolichothele	dŏ'lĭkōthē'lĕ
	Epithelantha	ĕ'pĭthĕlăn'thă
	Escobaria	ĕskōbă'rĭă
	Mammillaria	mămĭlā'rĭă
	Neolloydia	nēōloi'dĭă

Family. AMARYLLIDACEAE
ă'mărĭlĭdă'sĭē

| AGAVE | Agave | ăgā'hvĭ |
| | Manfreda | mănfrē'dă |

Family. BROMELIACEAE
brōmē'lĭāsĭē

| BROMELIA | Hechtia | hĕ'ktĭă |

Family. CRASSULACEAE
kră'sŭlāsĭē

| ECHEVERIA | Dudleya | dŭ'dlĭyă |
| | Graptopetalum | gră'ptōpĕ'tălŭm |

Family. EUPHORBIACEAE
yōofŏr'bĭāsĭē

| EUPHORBIA | Euphorbia | yōofo'rbĭă |
| | Jatropha | jătrō'fă |

Family. FOUQUIERIACEAE
fōokē'rĭāsĭē

| FOUQUIERIA | Fouquieria | fōokē'rĭă |

Family. LILIACEAE
lĭ′lĭāsĭē

YUCCA

Dasylirion	dă′sĭlĭ′rĭŏn
Hesperaloe	hĕspĕră′lō
Nolina	nōlē′nă
Yucca	yoŏ′kă

Family. PORTULACACEAE
po′rtŭlăkā′sĭē.

PORTULACA   Lewisia*   lōoĭ′zĭa

* *Note:* The following genera, *Acanthocereus* and *Lewisia*, although included in this simplified classication, are not described in detail as they are outside the scope of this book, but they are mentioned briefly in the text. The pronunciation of the names which is generally used is given in a phonetic form. The following key indicates the value of the vowel sounds which appear above. An acute accent indicates the stressed syllables.

KEY TO PRONUNCIATION

ā ē ī ō ū o͝o (mate, mete, mite, mote, mute, moot)
ă ĕ ĭ ŏ ŭ o͞o (rack, reck, rick, rock, ruck, rook)
ar er ir or ur (mare, mere, mire, more, mure)
ar er or (part, pert, port)
ah aw oi oor ow owr (bah, bowl, boil, boor, brow, bower)

12. SIMPLIFIED BOTANICAL DESCRIPTIONS OF SPECIES

A few species which do not come from the USA have been mentioned in passing but they are not described.

These species are arranged under families, in the same order as the classification list in the previous section. The plants described are in alphabetical order of generic and species names in each family. The abbreviated name following the plant name represents the Botanical Authority, that is the person who described it under that generic heading.

The better-known synonyms by which some of these plants are known are also listed, with the Botanical Authority who so named them. The habitat is given and the common name or names where known, also the plate number where illustrated.

Those who have difficulty in following the descriptions might start off with a species which has been illustrated and comparing the photograph with the description. We think that by doing this it will enable you to get a better idea of how to visualise another plant which is described. We have included at least one species from each genius represented in these regions of the USA, even if not illustrated, for example *Coloradoa*.

NOTE: var. = variety, used after the name of the species where there is a slight variation. fa. = forma is used in the case of a rather minor difference.

FAMILY: CACTACEAE

Ancistrocactus scheerii. B. & R.
ănsĭs'trŏkă'ktŭs shē'rĭē
GROUP: *Echinocactus.*
SYNONYMS: *Echinocactus scheerii.* Salm-Dyck.
Echinocactus brevihamatus. Engelm.
Echinocactus megarhizus. Rose.
Ancistrocactus brevihamatus. B. & R.
Ancistrocactus megarhizus. B. & R.

HABITAT: Duval, Frio, Hidalgo, Jim Wells, Kinney, Maverick, Starr, Val Verde, Webb and Zapata counties in Texas, also southwards into the Chihuahua region of Mexico. Among grass at low altitudes.

DESCRIPTION: Oval-shaped plant, usually solitary with up to 12 shallow ribs, but not as prominent as with some plants. Body of the plant bright green, but partially obscured by the brilliant golden-yellow spines. 2, sometimes 3, central spines, pointing upwards of which 1 is hooked and up to 1½ in. (3·8 cm) in length. Radials about 20, quite short, and almost pectinate in formation, pale yellow to grey. This information on the spination applies only to mature specimens; juvenile ones have few central spines, sometimes none. Flower 1 in. (2·5 cm) across, greenish-yellow. Fruit over 1 in. (2·5 cm) long, yellow to pink when ripe, scaly surface.

COMMON NAME: Fish-hook Cactus.
PLATE: No. 5.

Ariocarpus fissuratus. K. Schum.
ărĭŏkăr′pŭs fĭ′sūrăh′tŭs
GROUP: *Echinocactus.*
SYNONYMS: *Mammillaria fissurata.* Engelm.
Anhalonium fissuratum. Engelm.
Roseocactus fissuratus. Gerger.
Neogomesia fissurata. Castenada.

HABITAT: Brewster, Cameron, Maverick, Pecos, Presidio, Terrell, Val Verde and Webb counties in Texas, also in the Chihuahua and Coahuila regions of Mexico. Limestone desert or desert-scrub areas.

DESCRIPTION: Solitary, dwarf, globular plant, but with a completely flattened top surface, which is level with the surrounding ground, up to 5 in. (12·5 cm) in diameter. This flat top consists of a large number of flattened, triangular tubercles, the surfaces of which are crossed by grooves or fissures, hence the specific name. In habitat the surface of the plant, that is the warty tubercles, is brownish, but somewhat greener with cultivated plants. Areole at the base of each tubercle, full of white wool. No spines. Flowers appear from the centre of the plant, up to 2 in. (5 cm) in diameter,

standing up just above the plant, varying from pale pink to magenta in colour. Fruit oblong, about 1 in. (2·5 cm) long, off-white, but becoming dry and brown when ripe.

COMMON NAME: Living Rock, Star Rock and other names.
PLATES: Nos. 37 and 38.

Astrophytum asterias. Lem.
ă′strŏfē′tŭm ăstēr′ĭăs
GROUP: *Echinocactus.*
SYNONYM: *Echinocactus asterias.* Zucc.

HABITAT: Cameron and Starr counties in Texas, also in the Nuevo Leon and Tamaulipas regions of Mexico. Grassland and beneath bushes.

DESCRIPTION: Flattened, disc-shaped plant up to 6 in. (15 cm) in diameter, usually much less, and rarely more than 1½ in. (3·8 cm) high. It consists of 8 low ribs on which are placed the areoles containing pale-yellow to grey wool. No spines. Body of plant varies from dull green to olive coloured, with white scale-like marking scattered over it; more numerous between the ribs. Flower yellow with an orange-red throat up to 3 in. (7·5 cm) in diameter. Fruit oval ½ in. (1·3 cm) long, covered in wool; when ripe the fruit is dry so that it splits and the seeds can fall out.

COMMON NAME: Sea-Urchin Cactus or Sand Dollar.
PLATES: Nos. 8, 10 and 140.

Bergerocactus emoryi. B. & R.
be′rjĕrŏkă′ktŭs ĭmō′rĭē
GROUP: *Cereus.*
SYNONYM: *Cereus emoryi.* Engelm.

HABITAT: San Diego county, California, also Santa Catalina and San Clemente Islands off the coast of

Orange and San Diego counties, and further south into Baja California, Mexico. Coastal areas in chaparral country.

DESCRIPTION: Sprawling shrub, with branching slender stems up to 3 ft (1 m) in height and barely 2 in. (5 cm) in diameter. Spines dense, almost obscuring the green stems, which consist of up to 16 ribs. Up to 30 golden-yellow spines, of which up to 3 can be termed centrals, one of which pointed downwards, and up to 2 in. (5 cm) in length. Flower up to 1½ in. (3·8 cm) in diameter, yellow. Fruit, which forms later, is as densely spined as the plant itself, and the same colour.

COMMON NAME: Not known.
PLATES: Nos. 115 to 117.

Carnegiea gigantea. B. & R.
kărnē′gĭă jĭgăntē′ă
GROUP: *Cereus.*
SYNONYM: *Cereus giganteus.* Engelm.
Pilocereus engelmannii. Lemair.
Pilocereus giganteus. Rumpler.

HABITAT: Gila, Graham, Maricope, Mohave, Pima, Pinal and Yuma counties in Arizona also in eastern California and the northern Sonoran part of Mexico. Sandy desert and desert-scrub areas, below 4,000 ft (1,220 m).

DESCRIPTION: Large, free branching columnar cactus, up to 40 ft (12·2 m) in height. 12–30 ribs on a stem; each stem can vary from 12 in. (30 cm) to 20 in. (50 cm) in diameter. Spines up to 3 in. (7·5 cm) in length, the longest ones pointed downwards, others spreading, but not distinguishable between radials and centrals. These are white-grey, sometimes with a pink tinge when young. Flower up to 3 in. (7·5 cm) in diameter, waxy white colour, opening at night, but remaining open for much of the next day. Fruit up to 3 in. (7·5 cm) long and 1½ in. (3·8 cm) wide, red when ripe.

COMMON NAME: Saguaro Cactus.
PLATES: Nos. 67 to 72.

Coloradoa mesae-verdae. Boiss.
kŏlŏrădō′ă mă′să-vĕ′rdē
GROUP: *Echinocactus.*
SYNONYMS: *Echinocactus mesae-verdae.* Benson.
Sclerocactus mesae-verdae. Benson.

HABITAT: A very small area between Cortez, Colorado to near Shiprock, New Mexico, on dry hills and rocky areas, and just into Arizona.

DESCRIPTION: Small, globular, plant, solitary usually, occasionally 2–3 heads, each up to 3 in. (7·5 cm) in diameter and light green in colour. Tubercled, but the tubercles tend to form a small rib structure, often in a spiral, and where this is so up to 17 in number. 1 central spine (when present) erect, up to ¼ in. (6 mm) long, straw-yellow to grey. Radials up to 10, less than 1 in. (2·5 cm) long, same colour, flattened against the body of the plant. Flower 1½ in. (3·8 cm) in diameter, cream-yellow, with a brownish median stripe on each petal. Fruit nearly round, pale green in colour.

COMMON NAME: Mesa Verde Cactus.
NOT ILLUSTRATED

Coryphantha echinus. B. & R.
kŏrĭfă′nthă ĕ′kĭnŭs
GROUP: *Mammillaria.*
SYNONYMS: *Mammillaria echinus.* Engelm.

Mammillaria pectinata. Engelm.
Mammillaria radians var. *echinus*. K. Schum.
Coryphantha pectinata. B. & R.
Coryphantha radians var. *pectinata*. Borg.
Coryphantha cornifera var. *echinus*. Benson.

HABITAT: Brewster, Coke, El Paso, Howard, Pecos, Presidio, Terrell and Val Verde counties in Texas, also in the Chihuahua and Coahuila regions of Mexico. Limestone desert and desert-scrub areas.

DESCRIPTION: Dwarf globular to short cylindric plant, tubercled, but rather obscured by the spination. 3–4 central spines, straight or slightly curved, up to 1 in. (2·5 cm) long. Radials usually over 20, occasionally up to ⅓ in. (9 mm) long, usually less, somewhat pectinate, thinner than the centrals. Radials and centrals cream to pale yellow. Flower 3 in. (7·5 cm) across, yellow, and the anthers are orange, giving the appearance of an orange centre. Fruit oval, 1 in. (2·5 cm) long, green.

COMMON NAME: Hedgehog Cory-cactus
PLATES: Nos. 15 and 16.

Coryphantha muehlenpfordtii var. robustispina. Marshall.
kŏrĭfăn'thă mŏŏĕ'lĕnfŏr'dĭē rōbŏŏ'stĭspĕn'ă
GROUP: *Mammillaria*.
SYNONYMS: *Mammillaria robustispina*. Schott.
Coryphantha robustispina. B. & R.
Coryphantha scheerii var. *robustispina*. Benson.

HABITAT: Pima and Santa Cruz countries in Arizona, also in the northern Sonora region of Mexico, up to 5,000 ft (1,500 m), in desert or grassland areas, and even oak woodland occasionally.

DESCRIPTION: Globular to short cylindrical, up to 8 in. (20 cm) in height and as much as 4 in. (10 cm) in diameter. Prominently tubercled, these being quite large, ½ in. (1·25 cm) or more across at their bases, bearing at the conical tip the spine cluster. Usually one strong, hooked or curved central spine, grey coloured to horny-yellow. Up to 15 radials (but this number is far less in young specimens) spreading, less than 1 in. (2·5 cm) long, and of a similar colour to the central. Flower 1½ in. (4 cm) across, yellowish to pink. Fruit oblong 2 in. (5 cm) in length, green.

COMMON NAME: Robust Pincushion or Stout Needle Mulee.
NOT ILLUSTRATED

Coryphantha runyonii. B. & R.
kŏrĭfăn'thă rŭnyŏ'nĭē
GROUP: *Mammillaria*.
SYNONYMS: *Mammillaria runyonii*. Cory.
Lepidocoryphantha runyonii. Bckbg.
Coryphantha macromeris var. *runyonii*. Benson.

HABITAT: Starr county, Texas, and down into the Chihuahuan region of Mexico. Desert-scrub areas at low altitudes.

DESCRIPTION: Clustering, low-growing plant, prominently tubercled. Body of the plant rather soft, grey to blue-grey in colour. Rarely grows higher than 4 in. (10 cm), and each head rarely exceeds 2 in. (5 cm) in diameter. Up to 4 central spines, which sometimes reach 2 in. (5 cm) in length, brown in colour. The radials vary between 6 and 12, white to pale yellow in colour, barely ½ in. (1·3 cm) long, and more flattened against the plant compared with the centrals which stand quite erect. Flowers up to 2 in. (5 cm) in diameter, dark pink to magenta; some of the outermost petals have ciliate (fringed) edges. Fruit oval, over ½ in. (1·3 cm) long, green. This plant has quite a deep root system; many of the roots are swollen into a form of storage organ, although it could not be defined strictly as having a tuberous root system.

COMMON NAME: Big Nipple Cactus or Dumpling Cactus.

NOT ILLUSTRATED

Coryphantha vivipara.
var. **arizonica.** Marshall.
kŏrĭfă'nthă vĭvĭ'pără ărĭzŏ'nĭcă
GROUP: *Mammillaria.*
SYNONYMS: *Mammillaria arizonica.* Engelm.
Mammillaria vivipara var. *arizonica.* Benson.
Coryphantha arizonica. B. & R.
Escobaria arizonica. Buxbaum.

HABITAT: All the counties of Arizona except Santa Cruz and Yuma, also just into southern Nevada and Utah, and western New Mexico. On rocky ground, among trees between 4,000 and 7,000 ft (1,200 m and 2,300 m).

DESCRIPTION: Small globular species, solitary or forming into small clumps, each head rarely exceeding 3 in. (7·5 cm) in diameter. Up to 6 brownish central spines, under 1 in. (2·5 cm) in length. Up to 20 or so radial spines, white and shorter than the central spines. Flower up to 2 in. (5 cm) in diameter rose-pink colour. Outer parts of the flower, more equivalent to sepals, greenish-brown and with

fringed edges. Fruit oval, ¾ in. (2 cm) long, green.

COMMON NAME: Beehive Cactus.
PLATE: No. 107.

Coryphantha vivipara.
var. **bisbeeana.** Benson
kŏrĭfăn'thă vĭvĭ'pără bĭsbĭă'hnă
GROUP: *Mammillaria.*
SYNONYMS: *Coryphantha bisbeeana.* Orcutt.
Escobaria bisbeeana. Borg.
Mammillaria bisbeeana. Bckbg.
Coryphantha aggregata. B. & R.

HABITAT: Cochise, Pima, Pinal and Santa Cruz counties in Arizona, and in the nearby Sonora region of Mexico, on grassland and hillsides below 5,000 ft (1,600 m).

DESCRIPTION: Globular to slightly egg-shaped stems, usually freely branching, each about 2½ in. (6·4 cm) in diameter. Clumps can reach 2 ft (60 cm) or more in diameter in age. Usually 5 or 6 central spines, pinkish-grey to brown and about ½ in. (1·25 cm.) long, spreading so that they are not always easy to differentiate from the radials. Up to 30 radials, brown, sometimes white towards the base, slightly shorter than the centrals. Flower about 1½ in. (4 cm) in diameter, pink, petals being rather narrow. Fruit less than 1 in. (2·5 cm) long, green.

COMMON NAME: Hens and Chickens.
NOT ILLUSTRATED.

Dolichothele sphaerica. B. & R.
dŏlĭkŏthē'lĭ sfĕ'rĭkă
GROUP: *Mammillaria.*
SYNONYMS: *Mammillaria sphaerica.* Dietrich.
Mammillaria longimamma var. *sphaerica.* K. Brand.

HABITAT: Cameron, Duval, Kleberg, Nueces, San Patricio, Starr, Taylor and Val Verde counties in Texas, also in the nearby Chihuahua region of Mexico. Grassland and desert-scrub at low altitude.

DESCRIPTION: Dwarf, clumping plant, prominently tubercled, each head averaging around 2 in. (5 cm) in diameter. Up to 4 central spines, pale yellow to grey in age, occasionally brown about ½ in. (1·3 cm) long at the most. Up to 14 or 15 radials, similar colour to the centrals, tending to be pressed against the plant, about ⅓ in. (9 mm) long. Both types of spines are not stiff in any way. Flower yellow, often exceeding 2 in. (5 cm) in diameter, and noticeably fragrant. Fruit pale brownish-green when ripe and about ½ in. (1·3 cm) long.

COMMON NAME: Not known.

Note. This species is closely related to *D. longimamma* and *D. melaleuca* from Mexico, both of which have tubercles twice as long, and heads twice as large.

PLATE: No. 6.

Echinocactus horizonthalonius. Lem.

ĕkĭnōkă′ktŭs hŏ′rĭzŏntălō′nĭŭs

GROUP: *Echinocactus.*

SYNONYMS: *Echinocactus equitans.* Scheid. *Echinocactus laticostatus.* Engelm. *Echinocactus horizonthalonius* var. *curvispinus.* Engelm. *Echinocactus horizonthalonius* var. *nicholii.* Benson.

Also other variety names of *E. horizonthalonius.*

HABITAT: Brewster, Culberson, Duval, El Paso, Hudspeth, Jeff Davis, Pecos, Presidio, Terrell and Val Verde counties, in Texas, through the most southerly counties of New Mexico and into Cochise, Pima and Pina, counties in Arizona, as well as a wide area of northern Mexico bordering Texas and Arizona, and as far south as San Luis Potosi. Limestone desert and desert-scrub.

DESCRIPTION: Flattened globular to short cylindric plant, up to 8 in. (20 cm), usually solitary, but occasionally a clustered plant has been found. Usually 8 rounded ribs, body of the plant chalky-blue in colour, and very tough. Areoles full of white wool when young varying between ½ in. and 1 in. (1·3 and 2·5 cm) apart. 1 central spine strong, curved downwards, about 1 in. (2·5 cm) long. Radials up to 8, curved, spreading in all directions against the body of the plant, grey, slightly brown when young, as with the central. Flower varies from pale pink to a vivid rose-red, and up to 3 in. (7·5 cm) in diameter, opening wide. Fruit oblong, over 1 in. (2·5 cm) long, and ½ in. (1·3 cm) wide, reddish-brown when ripe, but often obscured by a mass of white wool in the crown of the plant.

COMMON NAME: Blue Barrel, Eagle's Claw and many other names.

Note. It is a very variable species, growing over a wide area, and we are convinced after growing it for some 40 years in the Exotic Collection, Worthing, England, that the columnar forms are just older mature specimens and not varieties.

PLATES: Nos. 34 and 36.

Echinocactus polycephalus.

Engelm & Bigelow.

ĕkĭnōkă′ktŭs pŏlĭsĕ′fălŭs

GROUP: *Echinocactus.*

SYNONYM: *Echinocactus polycephalus* var. *flavispinus*. Haage Jr.

HABITAT: Kern, Inyo, and San Bernardino counties in California, Mohave county, Arizona and just into Nevada and Utah, growing on rocky desert and hill areas, sometimes in woodland, at around 4,000 ft to 5,000 ft (1,220 m to 1,500 m).

DESCRIPTION: Usually a clumping species, each head being globular to short-cylindric, up to 24 in. (60 cm) high and 12 in. (30 cm) in diameter. A ribbed species, between 13 and 21, somewhat obscured by the dense spination. Up to 4 grey-red flattened central spines, sometimes exceeding 3 in. (7·5 cm) in length, but often possessing a soft felt-like layer which peels off. Radials, up to 8 per areole, spreading, also slightly flattened, and about 1½ in. (3·8 cm) long. Flower 2 in. (5 cm) in diameter, yellow, sometimes with a pink median stripe on each petal. Fruit globular, covered with white wool.

COMMON NAME: Woolly-headed Barrel Cactus.

NOT ILLUSTRATED.

Echinocereus dasyacanthus.
Engelm.
ĕkĭnōsē′rĭŭs dă′sĭăkă′nthŭs
GROUP: *Cereus.*

SYNONYM: *Cereus dasyacanthus*. Engelm. *Cereus dasyacanthus* var. *neomexicanus*. Coulter.
Echinocereus steerae. Glover.
Echinocereus pectinatus var. *steerae*. Rumpler.
Echinocereus dasyacanthus var. *steerae*. Bckbg.
Echinocereus pectinatus var. *neomexicanus*. Benson.

HABITAT: Brewster, Culberson, Hudspeth, Mitchell, Pecos, Presidio, Terrell and Upton counties in Texas, most of the southern counties of New Mexico, into Arizona in Cochise, Pima, and Santa Cruz counties, and over a wide area of the northern Chihuahua region of Mexico. Limestone desert and desert-scrub below 4,000 ft (1,220 m).

DESCRIPTION: Cylindrical stemmed plant, solitary when young, but clustering in age, occasionally reaching 16 in. (40 cm) or so in height, while the stems can be up to 3 in. or 4 in. (7·5 cm and 10 cm) in diameter. Up to 20 narrow ribs, the areoles about ½ in. (1·3 cm) or less apart. Between 3 and 5 strong central spines, occasionally more, spreading between ½ in. and 1 in. (1·3 cm and 2·5 cm) long. Radials about 20, spreading, about ½ in. (1·3 cm) long. The spine colour can be very variable, from off-white, through yellow, and brown to some with quite vivid red tips. Flower up to 5 in. (12·5 cm) in diameter, occasionally even larger, particularly in cultivation. Petals are rather ragged at the tips, always green at the base, but the rest of the petal can vary from shades of yellow to pink or red. Fruit globular, up to 2 in. (5 cm) in diameter, spiny, reddish-brown, when ripe.

COMMON NAME: Texas Rainbow Cactus, and other names.

PLATES: Nos. 46 and 47.

Note: Reddish-spined form shown in flower, left centre.

Echinocereus baileyi. B. & R.
ĕkĭnōsē′rĭŭs bā′lĭē
GROUP: *Cereus.*

SYNONYMS: *Echinocereus caespitosus.*
Bckbg.
Echinocereus baileyi var. *caespitosus.*
Bckbg.
Echinocereus mariae. Bckbg.

HABITAT: Granite-covered slopes in the Wichita Mountains, Oklahoma.

DESCRIPTION: A short cylindrical plant, sometimes solitary, but more often forming into quite large groups, single stems up to 8 in. (20 cm) high and 2–3 in. (5–7·5 cm) in diameter. 12–15 shallow ribs on average, green, with areoles quite close together, these having some wool. Up to 5 short central spines, sometimes none, but when present rarely exceeding ¼ in. (6 mm) in length. Between 15 and 28 spreading radial spines up to a maximum of 1 in. (2·5 cm) long. Spine colouration is very variable from colony to colony, white or yellow to reddish-brown. Flower up to 3 in. (7·5 cm) in diameter, opening wide, varying shades of magenta to purple, with a darker centre, whilst the petal shape is varied even in one flower, as some have cut ends which gives a very ragged appearance. Fruit spiny, egg shaped, just under 1 in. (2·5 cm) long, olive-green.

COMMON NAME: Not known.
PLATE: No. 139.

Echinocereus dubius. Rumpler.

ĕkĭnŏsē′rĭŭs dūbĭŭs
GROUP: *Cereus.*
SYNONYMS: *Cereus dubius.* Engelm.
Echinocereus enneacanthus var. *dubius.* Benson.

HABITAT: Brewster, Culberson, Hudspeth, Presidio and Val Verde counties in Texas, and the nearby Chihuahua regions of Mexico.

Limestone desert at low altitudes.
DESCRIPTION: A somewhat sprawling species, but the ends of the rather soft stems are held more erect, occasionally up to 20 in. (50 cm) in length, and up to 3½ in. (8·8 cm) in diameter. Between 7 and 10 ribs, areoles circular and up to ¼ in. (7 mm) in diameter, and over 1 in. (2·5 cm) apart, bearing some white wool when young. Spines off-white usually 1, but occasionally more curved, centrals up to 3 in. (7·5 cm) long. Radials, between 5 and 10, thin, spreading, variable in length from ½ in. to 1½ in. (1·3 cm to 3·8 cm). Flower about 3 in. (7·5 cm) in diameter, magenta-pink. Fruit globular, just over 1 in. (2·5 cm) in diameter, spiny, reddish-brown to purple when ripe.

COMMON NAME: Purple Pitaya.
PLATES: Nos. 53 and 54.

Echinocereus engelmannii. Lem.

ĕkĭnŏsē′rĭŭs ĕngĕlmă′nĭē
GROUP: *Cereus.*
SYNONYMS: *Cereus engelmannii.* Parry ex. Rumpler.
Cereus engelmannii var. *variegatus.* Engelm.
Echinocereus engelmannii var. *variegatus.* Engelm.

HABITAT: Maricopa, Mohave, Pima and Yuma counties in Arizona, also in the four counties of California bordering with Arizona, and San Diego county, and down in the Baja California and northern Sonora parts of Mexico. Sandy and rocky areas below 5,000 ft (1,600 m).

DESCRIPTION: Low growing clustering cactus with up to twelve or more erect stems, which sometimes reach 10 in. (25 cm) in height and average around

2 in. (5 cm) in diameter. Up to 5 central spines 2 in. (5 cm) or more in length, but usually only 1 of the 5 is particularly long. Radial spines rarely more than $\frac{1}{2}$ in. (1·3 cm) in length, and up to 12 in number. Spine colouration very variable from white to yellow or even pink. Flower about $2\frac{1}{2}$ in. (6 cm) in diameter in varying shades of magenta. Fruit round, usually over 1 in. (2·5 cm) in diameter, red when ripe, spiny.

COMMON NAME: Strawberry Hedgehog Cactus.

PLATES: Nos. 92, 93, 111 and 113.

Echinocereus engelmannii. var. **nicholii.** Benson.

ĕkĭnōsē'rĭŭs ĕngĕlmă'nĭē nĭkŏ'lē

GROUP: *Cereus.*

SYNONYMS: None.

HABITAT: Pima and Yuma counties in Arizona, also in northern Sonoran part of Mexico. Desert and rocky hillsides below 3,000 ft (1,000 m).

DESCRIPTION: Low-growing clustering variety, but compared with *E. engelmannii,* differs in the following respects. Forms into clumps of thirty or more stems, each stem sometimes attaining 24 in. or more in height (over 60 cm), and up to 3 in. (7·5 cm) in diameter. Spines usually a golden-yellow colour, and the centrals slightly longer. Flower colour usually described as much paler, but quite variable, and similar in size to *E. engelmannii.* Fruit similar to that of *E. engelmannii.*

COMMON NAME: Nichol's Hedgehog Cactus.

Note. The dark flowered specimen with the longer centred spines illustrated in Plate 75 was from the Silver Bells, west of Tucson, Arizona.

PLATES: Nos. 75 and 76.

Echinocereus enneacanthus. Engelm.

ĕkĭnōsē'rĭŭs ĕ'nĭăkă'nthŭs

GROUP: *Cereus.*

SYNONYM: *Cereus enneacanthus.* Engelm. *Echinocereus enneacanthus* var. *brevispinus.* Benson.

HABITAT: Brewster, Cameron, El Paso, Hays, Hidalgo, Kerr, Kimble, Kinney, Maverick, McMullen, Pecos, Presidio, Starr, Terrell, Tom Green, Val Verde, Webb and Wells counties in Texas, also into southern New Mexico, and the Chihuahua, Coahuila and Nuevo Leon regions of Mexico. Desert and desert-scrub areas below 3,000 ft (1,000 m).

DESCRIPTION: Sprawling, freely branching plant, ends of the branches usually holding their heads erect. Stems up to 30 in. (75 cm) in length, occasionally and usually 2 in. (5 cm) in diameter, sometimes more with the very robust forms, green in colour. Up to 10 ribs, but 7–8 is more usual, areoles up to 1 in. (2·5 cm), apart. 1 central spine, occasionally 3, up to 2 in. (5 cm) long, straw-coloured. Radials can number up to 12, but 8 is more common, about $\frac{1}{2}$ in. (1·3 cm) long, and of a similar colour to the centrals. Flower opening wide, 3 in. (7·5 cm) in diameter, varying shades of magenta-pink. Fruit round, 1 in. (2·5 cm) in diameter, greenish-brown, spiny.

COMMON NAME: Strawberry Cactus or Pitaya.

Note. Stems of this species in its many forms often tend to remain somewhat wrinkled, also the spine clusters are easily knocked off.

NOT ILLUSTRATED.

Echinocereus fendleri. Lem.
var. **bonkerae.** Benson.
ĕkĭnōsē'rĭŭs fĕ'ndlĕrē bŏ'nkĕrăē
GROUP: *Cereus.*
SYNONYMS: *Echinocereus bonkerae.* Thornber.
Echinocereus boyce-thompsonii var. *bonkerae.* Peebles.
Echinocereus fasciculatus var. *bonkerae.* Benson.

HABITAT: Cochise, Gila, Maricope, Pinal and Santa Cruz counties, Arizona, on sandy hillsides, grassland, and woodland, below 6,000 ft (2,000 m).
DESCRIPTION: Low growing clumping plants, with up to 10 or more cylindrical erect stems, up to 20 in. (50 cm) long, but usually half that length, and up to 3 in. (7·5 cm) in diameter. Up to 14 or so ribs, dark green body, areoles fairly close-together. Up to 3 erect ½ in. (1·3 cm) long central spines. Radials vary from 9–12, spreading, also about ½ in. (1·3 cm) long, straw-yellow to very dark brown, occasionally pure white, the same as the centrals. Flower up to 3 in. (7·5 cm) in diameter, varying shades of magenta. Fruit somewhat spiny, nearly round, red.
COMMON NAME: Bonker's Hedgehog Cactus.
PLATES: Nos. 123 and 124.

Echinocereus fendleri.
var. **robustus.** Benson
ĕkĭnōsēr'ĭŭs fĕ'ndlĕrē rōbŏō'stŭs
GROUP: *Cereus.*
SYNONYMS: *Echinocereus rectispinus* var. *robustus.* Peebles.
Echinocereus robustus. Peebles.
Echinocereus fasciculatus. Benson.

HABITAT: Cochise, Pima, Pinal and

Santa Cruz countries in Arizona, and in Mexico in the Chihuahua and Sonora regions of Mexico, growing in quite varied locations from in rocky and sandy areas as well as in grassland up to 4,000 ft (1,300 m).
DESCRIPTION: Low growing clumping plants with up to 20 stems, these being cylindrical up to 18 in. (60 cm) long and 3 in. (7·5 cm) in diameter with 8–10 ribs, occasionally more. Circular areole usually bears one central spine, standing out from the plant, dark-brown, from 1½ in (4 cm) to 3 in. (7·5 cm) long. About 12 radials, paler in colour, even white or grey occasionally, spreading and up to ¾ in. (2 cm) long. Flower about 2½ in. (6 cm) in diameter, shades of purple and magenta. Fruit nearly round, about 1 in. (2·5 cm) in diameter, purplish red, spiny.
COMMON NAME: Robust Hedgehog.
NOT ILLUSTRATED.

Echinocereus fendleri. Lem.
var. **boyce-thompsonii.** Benson.
ĕkĭnōsē'rĭŭs fĕn'dlĕrē bŏ'ĭstŏ'mpsŏnĭē
GROUP: *Cereus.*
SYNONYMS: *Echinocereus boyce-thompsonii.* Orcutt.
Echinocereus fasciculatus var. *boyce-thompsonii.* Benson.

HABITAT: Gila, Maricopa, Pinal and Yavapai counties in Arizona. Desert and desert-scrub areas below 3,000 ft (1,000 m).
DESCRIPTION: Low-growing clumping plants with 10 or more erect stems, sometimes reaching 10 in. (25 cm) in height and usually around 2½ in. (6 cm) in diameter. 14–16 ribs densely spined, the spines being very variable

in colour from brown, to some nearly white, others almost black in a few cases. Up to 4 central spines, 1 of which can be 2 in. (5 cm) in length, and up to 10 spreading radial spines which are somewhat shorter. Flower in varying shades of purple or magenta and around 2½ in. (6 cm) in diameter. Fruit nearly round 1 in. (2·5 cm) in diameter, red when ripe, spiny.

COMMON NAME: Boyce–Thompson Hedgehog Cactus.

PLATES: Nos. 97, 98 and 99.

Echinocereus fitchii. B. & R.

ĕkĭnōse'rĭŭs fĭ'tchĭē

GROUP: *Cereus.*

SYNONYM: *Echinocereus reichenbachii* var. *fitchii.* Benson.

HABITAT: Starr, Webb and Zapata counties in Texas and also just into the Chihuahua region of Mexico to the south. Among grass and desert-scrub areas at low altitude.

DESCRIPTION: Usually solitary, short columnar plant, occasionally with a few branches, stems rarely exceeding 6 in. (15 cm) in height, 2½ in. (6 cm) diameter, and consisting of up to 14 ribs. Up to 7 thin central spines, ¼ in. (6 mm) long, spreading in 2 rows outwards, over the radials, which are usually 20 in number, of a similar length, also very thin. The colour of these spines can vary a great deal from light brown to reddish-brown for their entire length, some with black tips. Flower pink-magenta and up to 4 in. (10 cm) in diameter, opening wide. Fruit up to 1 in. (2·5 cm) long, spiny and, as with the plant, almost obscuring the green.

COMMON NAME: Not known.

NOT ILLUSTRATED.

Echinocereus melanocentrus.

Lowry.

ĕkĭnōsē'rĭŭs mĕlănōsĕn'trŭs

GROUP: *Cereus.*

SYNONYM: *Echinocereus reichenbachii* var. *albertii.* Benson.

HABITAT: Jim Wells and Kleberg counties, in grassland areas, but usually growing beneath bushes.

DESCRIPTION: Clustering species made up of numerous short cylindrical stems, occasionally up to 6 in. (15 cm) long and 2 in. (5 cm) in diameter. Between 10 and 13 low ribs, body dark blue-green. Areoles very close together. 1 black central spine, less than ¼ in. (6 mm) long, not always present. Up to 20 radial spines, pectinate, mostly the same or shorter than the central spine, but off-white. Flower up to 3 in. (7·5 cm) in diameter, rose-magenta, more vivid in the centre, opening wide with many recurved petals. Fruit spiny, egg-shaped, ½ in. (1·3 cm) long, dark blue-green.

COMMON NAME: Not known.

PLATE: No. 140.

Echinocereus pentalophus.

Rumpl.

ĕkĭnōsē'rĭŭs pĕntălŏ'fŭs

GROUP: *Cereus.*

SYNONYMS: *Cereus pentalophus* DC
Cereus leptacanthus DC
Cereus procumbens. Engelm.
Echinocereus leptacanthus. K. Schum.
Echinocereus procumbens. Rumpler.

HABITAT: Bexar, Cameron, Duval, Hidalgo, Starr and Webb counties in Texas, and just over the Rio Grande River into Mexico. Among grass and desert-scrub at low altitudes.

DESCRIPTION: Low growing, crawling

habit, soft light green stems up to 12 in. (30 cm) in length, up to a maximum of 1 in. (2·5 cm) diameter, 4–5 ribs, free branching. Central spines rarely present, but if so similar to the radials, up to ½ in. (1·3 cm) long, pale brown. Up to 6 radial spines, spreading in different directions. Flower up to 4 in. (10 cm) in diameter, pink with a pale centre, rounded petals. Fruit oval, green, spiny, just under 1 in. (2·5 cm) long.

COMMON NAME: Lady Finger Cactus.

NOT ILLUSTRATED.

Echinocereus rigidissimus. Rose.

ĕkĭnōsē′rĭŭs rĭjĭdĭ′sĭmŭs

GROUP: *Cereus.*

SYNONYMS: *Cereus pectinatus.* Scheid.
Echinocereus dasyacanthus var. *rigidissimus.* Mshll.
Echinocereus pectinatus var. *rigidissimus.* Benson.

HABITAT: Cochise, Pima and Santa Cruz counties in Arizona. Also in New Mexico and northern Sonora in Mexico. Among grass and on limestone hills at around 4,000 ft (1,300 m).

DESCRIPTION: Simple, erect, columnar, cactus rarely exceeding 12 in. (30 cm) in height. Short pectinate radial spines, varying from white to red, often forming a colourful banding effect on the stem. No central spines. Flower about 3 in. (7·5 cm) in diameter, purple to red with a white throat. Fruit oval nearly 2 in. (5 cm) long, purplish-red when ripe, spiny.

COMMON NAME: Arizona Rainbow Hedgehog.

PLATES: Nos. 56 to 59.

Echinocereus stramineus. Engelm.

ĕkĭnōsērĭŭs strămĭ′nĭŭs

GROUP: *Cereus.*

SYNONYMS: *Cereus stramineus.* Engelm.
Echinocereus bolansis. Runge.
Echinocereus stramineus var. *major.* Hort.
Echinocereus enneacanthus var. *stramineus.* Benson.

HABITAT: Brewster, Culberson, El Paso, Jeff Davis, Pecor, Presidio and Terrell counties in Texas. Limestone desert and desert-scrub below 4,000 ft (1,300 m).

DESCRIPTION: A plant that can grow into huge mounds, of 300 or more stems. Single stems can be 10 in. (25 cm) high and over 3 in. (7·5 cm) in diameter, green, and consisting of about 12 ribs. Areoles small, white, bearing off-white to straw yellow spines. Up to 4 central spines, erect, sometimes exceeding 3 in. (7·5 cm) in length. The radials vary between 10 and 14, and are sometimes over 1 in. (2·5 cm) in length, spreading in a way which tends to obscure the body of the stem. Flower up to 5 in. (12·5 cm) in diameter, and nearly as tall, petals often somewhat ragged or toothed at the tips, dark rose to varying shades of magenta. Fruit round, up to 2 in. (5 cm) in diameter, red when ripe, covered in quite long bristle-like spines.

COMMON NAME: Strawberry Cactus or Pitaya.

PLATES: Nos. 27 to 30.

Echinocereus triglochidiatus.

ĕkĭnōsē′rĭŭs trĭglŏkĭdĭā′htŭs Engelm.

GROUP: *Cereus.*

SYNONYMS: *Cereus triglochidiatus.* Engelm.

Echinocereus paucispinus var. *triglochidiatus.* Schumann.

Note. This is a very confused species, and other synonyms could be included depending on how many varieties of *triglochidiatus* are acceptable.

HABITAT: Widespread distribution through much of the north and western parts of Texas, northern New Mexico, also in Arizona, Colorado, etc. Hillsides among grass and trees around 4,000 ft–7,000 ft (1,300 m–2,300 m).

DESCRIPTION: Low-growing clumps, made up of a dozen or more stems, up to 8 in. (20 cm) high and often 3 in. (7·5 cm) in diameter. 6–8 ribs, the green stems often tend to appear rather wrinkled, particularly in the resting period. Areoles have a little white wool, from which the grey to black spines appear. The spines, which are all radials and up to 2 in. (5 cm) or so in length, have a bulbous base and are usually somewhat flattened. 3–6 spines is the more usual, although the forms from more northerly localities invariably have 3. Flower with oval shaped petals quite waxy, varying from orange-red to scarlet, and around 1½ in. (3·8 cm) across. Fruit around 1 in. (2·5 cm) diameter, purplish-red, spiny.

COMMON NAME: Claret Cup Hedgehog. NOT ILLUSTRATED.

Echinocereus triglochidiatus var. octacanthus. Marshall.

ĕkĭnōsē′rĭŭs trĭglŏkĭdĭā′htŭs ŏktăkă′nthŭs

GROUP: *Cereus.*

SYNONYMS: *Cereus roemeri.* Engelm. *Cereus octacanthus.* Coulter. *Echinocereus roemeri.* Rumpler.

Echinocereus octacanthus. B. & R. *Echinocereus paucispinus.* Engelm. *Echinocereus triglochidiatus* var. *paucispinus.* Marshall.

HABITAT: Jeff Davis, Kimble, Pecos, Presidio, and Val Verde counties in Texas, also into south-eastern New Mexico and in Mexico bordering the Big Bend area of Texas. Desert and desert-scrub areas below 4,000 ft (1,300 m).

DESCRIPTION: Clustering species, sometimes into clumps of a dozen or so stems which are erect and up to 10 in. (25 cm) or so in height, and up to 4 in. (10 cm) in diameter. Up to 9 or so ribs. Up to 9 radial spines, all rounded not flattened as with *E. triglochidiatus*, and often 1 central spine, erect, sometimes exceeding 1 in. (2·5 cm) in length. Spine colour varies from off-white to occasionally even black. Flower 1½ in. (3·8 cm) in diameter in shades of orange-red. Fruit similar to *E. triglochidiatus*.

COMMON NAME: Claret Cup Hedgehog. PLATES: Nos. 18 and 19.

Echinocereus triglochidiatus var. melanacanthus. Benson.

ĕkĭnōsē′rĭŭs trĭglŏkĭdĭā′htŭs mĕlănăkā′nthŭs

GROUP: *Cereus.*

SYNONYMS: *Echinocereus melanacanthus.* Engelm.

Echinocereus coccineus var. *melanacanthus.* Engelm.

Another possible twelve synonyms could be listed as referring to this variety.

HABITAT: All the counties of Arizona, at the higher altitudes, also in parts of California, Nevada, Utah, Colorado, New Mexico, Texas, and down into

Mexico for quite a distance. Rocky hillsides, among grass and trees at between 4,000 ft–10,000 ft (1,300 m–3,300 m).

DESCRIPTION: Forms into very large clumps or mounds of up to 500 heads, with single stems up to 6 in. (15 cm) in length, and 2½ in. (6 cm) in diameter. Up to 9 ribs, up to 3 white 2½ in. (6 cm) long central spines. Up to 10, occasionally more radial spines, about 1 in. (2·5 cm) long and of a similar colour to the centrals. Flower usually orange-red and about 1½ in. (3·8 cm) in diameter. Fruit similar to *E. triglochidiatus*.

Note. The specific name *melanacanthus* refers to black spines, whereas in fact grey or black spined plants are far less common than the white spined ones.

COMMON NAME: White Spined Claret Cup Hedgehog.

PLATES: Nos. 103 and 104.

Echinocereus viridiflorus. Engelm.
ĕkĭnōsē'rĭŭs vĭrĭdĭflō'rŭs

GROUP: *Cereus.*

SYNONYMS: *Cereus viridiflorus.* Engelm.
Cereus viridiflorus var. *minor.* Engelm.
Echinocereus standleyi. B. & R.

HABITAT: Over a very wide area from near the Rocky Mountains, through Wyoming, Colorado, New Mexico, South Dakota, into Texas, and just into Arizona. Grows in stony areas, in grassland and woodland, below 8,000 ft (2,600 m). The small growing northerly forms and those from higher altitudes stand severe winter conditions if in a well-drained position.

DESCRIPTION: Solitary, but more often clustering, species, made up of short cylindrical stems between 1 in. and 3 in. (2·5–7·5 cm) high and up to 1½ in.

(3·8 cm) in diameter. Distinctly ribbed structure, between 12 and 15 ribs, green or greenish-brown body. Radial spines only, flattened against the body of the plant, between 10 and 20, up to ½ in. (1·3 cm) long, but often half this length, and even shorter still with the smaller growing forms of this species. Spines vary from white, yellow through to brown or even red. Flower ½ in.–1 in. (1·3–2·5 cm) in diameter, not opening very wide in some cases, varying shades of green or yellowish green, strongly lemon scented. Fruit spiny, egg shaped up to ½ in. (1·3 cm) long, greenish-brown.

COMMON NAME: Green-flowered Torch Cactus, Green-Flowered Pitaya and many other names.

PLATE: No. 125.

Echinomastus johnsonii. Baxter.
ĕkĭnōmă'stŭs jŏ'nsŏnē

GROUP: *Echinocactus.*

SYNONYMS: *Echinocactus johnsonii.* Parry.
Ferocactus johnsonii. B. & R.
Echinomastus arizonicus. Hester.
Neolloydia johnsonii. Benson.

Note. The variety name *lutescens* also appears, but this is basically a colour variation of the flower.

HABITAT: Mohave, Yavapai and Yuma counties in Arizona, Inyo and San Bernadino counties in California, and just extending into Nevada and Utah. Desert and rocky hillsides up to 4,000 ft (1,300 m).

DESCRIPTION: Solitary, globular to short cylindric plant, up to 10 in. (25 cm) high and 4 in. (10 cm) in diameter. Very densely spined, the spines pink to red, obscuring the body of the plant, become dirty-coloured when old. Up to 8 central spines,

H

some reach $1\frac{1}{2}$ in. (3·8 cm) long, erect. Radials pale coloured, sometimes cream coloured, spreading, up to 10, about $\frac{3}{4}$ in. (2 cm) long. Flower up to 3 in. (7·5 cm) in diameter, but rarely able to open properly because of the crowded spines at the apex of the plant, greenish-yellow or pink. Fruit oval, $\frac{1}{2}$ in. (1·3 cm) long, brown when ripe, dry.

COMMON NAME: Johnson's Pineapple Cactus.

Note. The greenish-yellow flowered form illustrated is the one that went under the variety name *lutescens*, and is more commonly found in association with *Yucca brevifolia* (Joshua Tree).

PLATE: No. 122.

Epithelantha bokei. Benson.
ĕpĭthĕlă'nthă bŏ'kĭē
GROUP: *Mammillaria*.

SYNONYM: None, other than plants being wrongly identified, and placed under *Epithelantha micromeris* (Weber) or the var. *greggii*. Engelm.

HABITAT: Brewster county in Texas. Limestone desert and desert-scrub below 4,000 ft (1,3000 m).

DESCRIPTION: Dwarf, almost globular, plant, solitary usually, occasionally 3–4 heads up to 2 in. (5 cm) in diameter, but often much less, somewhat flat topped. Tubercles very small and close together. Areoles bear small white spines, under $\frac{1}{5}$ in. (5 mm) in length, all of them flattened against the body of the plant. The spines are arranged in 4 or 5 series or layers, varying in number from 28 on the outer layer or series to 10 on the inner. Flower just under $\frac{1}{2}$ in. (1·3 cm) in diameter, off-white to pale pink in colour. Fruit cylindrical to club

shaped, about $\frac{1}{2}$ in. (1·3 cm) long, red when ripe.

COMMON NAME: Button Cactus.
PLATES: Nos. 39 and 40.

Epithelantha micromeris. Weber.
ĕpĭthĕlă'nthă mĭ'crŏmĕ'rĭs
GROUP: *Mammillaria*.
SYNONYMS: *Mammillaria micromeris*. Engelm.
Cactus micromeris. Kuntze.

HABITAT: Bandera, Brewster, Culberson, El Paso, Pecos, Presidio and Val Verde counties in Texas, also in all the southernmost counties in Arizona, and in the neighbouring regions of Mexico. Limestone desert and desertscrub below 4,000 ft (1,300 m).

DESCRIPTION: Dwarf, almost globular, plant, sometimes solitary, but often forming into large clumps, each head can occasionally exceed 2 in. (5 cm) in diameter, but usually much less, also more oval shaped. Tubercles very small and close together. Areole bearing white or off-white spines, nearly $\frac{1}{4}$ in. (6 mm) long, flattened against the body of the plant. The spines are in 2 or 3 series or layers, up to 20 in number, and sometimes accompanied by small white hairs, also coming from the areole. Flower about $\frac{1}{5}$ in. (5 mm) in diameter, pale pink, and sometimes only just pushing out from among the spines in the centre of the plant. This appears to occur with specimens that have been flowering for many years, rather than the young one illustrated (Plate 41). Fruit about $\frac{1}{2}$ in. (1·3 cm) long, club-shaped or cylindrical, red when ripe.

COMMON NAME: Button Cactus.
Note. *E. bokei* was only described by L. Benson in 1969. Although there has been some confusion over this plant

for many years, we are convinced that it is a good new species.

PLATE: No. 41.

Escobaria runyonii. B. & R.

ĕscōbă′rĭă rŭ′nyŏnĭē

GROUP: *Mammillaria.*

SYNONYMS: *Mammillaria escobaria.* Cory. *Mammillaria robertii.* Wngr. *Coryphantha robertii.* Bgr.

HABITAT: Duval, Hidalgo, Starr, Val Verde, Webb and Zapata counties in Texas, also in the Coahuila and Tamaulipas regions of Mexico. Among grass and on limestone desert-scrub areas below 1,000 ft (300 m).

DESCRIPTION: Dwarf globular plant, forming into large clumps, when each head tends to become somewhat elongated, about 2 in. (5 cm) high and 1 in. (2·5 cm) in diameter at the widest point. Densely tubercled, but tubercles are small, completely obscured by the spines, up to 8 thin centrals, brown with a black tip, and under ½ in. (1·3 cm) long. Radials up to 30, white, ¼ in. (6 mm) long, and tending to spread against the body of the plant. Flower just over ½ in. (1·3 cm) in diameter, cream with a dull brown median stripe on some of the petals. Fruit small, round, red when ripe.

COMMON NAME: Junior Tom Thumb Cactus.

PLATE: No. 4.

Escobaria sneedii. B. & R.

ĕscōbă′rĭă snē′dĭē

GROUP: *Mammillaria.*

SYNONYM: *Mammillaria sneedii.* B. & R.

HABITAT: Franklin Mountains, near El Paso, Texas and Las Cruces, New Mexico.

DESCRIPTION: Small, clustering species, each head or stem somewhat egg shaped, up to 2 in. (5 cm) long and just under 1 in. (2·5 cm) in diameter, prominently tubercled. 10–17 radiating central spines, slightly thicker than the radials, up to ¼ in. (6 mm) long. Radials, very fine, between 25 and 40, very short, flattened against the body, white as are the centrals. Flower not wide opening ½ in. (1·3 cm) long, white to pale pink, some of the outer petals with fringed edges. Fruit globular ¼ in. (6 mm) pink.

COMMON NAME: Not known.

Note. There is another closely related species *Escobaria leei* (syn *Mammillaria leei*) coming from near Carlsbad, New Mexico. It has wide opening flowers, little larger, and the plant is even more densely spined. We feel rather doubtful that these can be kept much longer as distinct species, because of their similarities, and the close proximity in nature.

NOT ILLUSTRATED.

Escobaria tuberculosa. B. & R.

ĕscōbă′rĭă tūbĕ′rcŭlō′să

GROUP: *Mammillaria.*

SYNONYMS: *Mammillaria strobiliformis.* Scheer. *Echinocactus strobiliformis.* Poselger. *Mammillaria tuberculosa.* Engelm. *Escobaria strobiliformis.* Scheer. *Coryphantha strobiliformis.* Orcutt.

HABITAT: Brewster, El Paso, Presidio and Terrell counties in Texas, also in the Chihuahua region of Mexico. Limestone desert and desert-scrub below 4,500 ft (1,500 m).

DESCRIPTION: Globular to short-cylindric plant, solitary or forming into small clumps. Single stems occasionally exceeding 6 in. (15 cm) in height, but often much less, and about 2 in. (5 cm) in diameter. Prominently tubercled, as the name suggests, body of the plant grey-green, becoming grey to brown in age, when the spines are often missing from the bottoms of stems. A little white wool on the areoles, near the top of a stem. Central spines 6–9 and up to $\frac{1}{2}$ in. (1·3 cm) long, grey in colour, often with pink tip. Radials over 20, sometimes as many as 30, white or off-white, spreading out in all directions, against the stem. Flower just over 1 in. (2·5 cm) in diameter, off-white to pink, opening wide. Fruit oblong, under 1 in. (2·5 cm), red when ripe.

COMMON NAME: Cob Cactus.

PLATE: No. 27.

Ferocactus acanthodes. B. & R.

fĕ'rōkăktŭs ăkănthō'dēs

GROUP: *Echinocactus*.

SYNONYMS: *Echinocactus acanthodes*. Lem. *Echinocactus viridescens* var. *cylindraceus*. Engelm.

Ferocactus rostii. B. & R.

HABITAT: Coconino, Gila, Graham, Maricopa, Mojave, Pima, Pinal and Yuma counties in Arizona, also in adjoining California, Baja California and northern Sonoran parts of Mexico. In sandy desert areas and on rocky hillsides below 3,000 ft (1,000 m).

DESCRIPTION: Usually solitary, globular, becoming columnar in age, up to 10 ft (3·3 m) in height and can exceed 12 in. (30 cm) in diameter. Tubercled when young, then changing to a ribbed structure, which can average around 24. 4 curved central spines in a cross formation up to 5 in. (12·5 cm) in length and up to 12 radials, tending

to flatten against the plant usually, around 2 in. (5 cm) in length. The spine colour is very variable, from cream to red. Flower yellow, up to 2 in. (5 cm) in diameter, but rarely able to open properly, because of the spination. Fruit oval, up to 1$\frac{1}{2}$ in. (3·8 cm) long and under 1 in. (2·5 cm) in diameter. Yellow when ripe.

COMMON NAME: Compass or California Barrel Cactus.

PLATES: Nos. 73, 74 and 95.

Ferocactus covillei. B. & R.

fĕ'rōkăktŭs kōv'illiē

GROUP: *Echinocactus*.

SYNONYM: *Echinocactus covillei*. Bgr.

HABITAT: Pima and Yuma counties in Arizona, also in northern Sonoran part of Mexico. Sandy desert and rocky areas below 3,000 ft (1,000 m).

DESCRIPTION: Usually solitary, globular to short cylindric cactus, only occasionally reaching 5 ft (1·6 m) in height. Tubercled when young, later forming the usual rib structure. 1 long hooked central spine up to 5 in. (12·5 cm) long, red when young, becoming grey later. Radial spines varying, 6–9 on average, about 2$\frac{1}{2}$ in. (6 cm) long, curved, but somewhat flattened against the plant, and of a similar colour to the central spines. Flower 3 in. (7·5 cm) in diameter, purplish red. Oval about 1$\frac{1}{2}$ in. (3·8 cm) long, yellow.

COMMON NAME: Coville's Barrel Cactus.

PLATES: Nos. 65 and 66

Ferocactus viridescens. B. & R.

fĕ'rōkăktŭs vĭrĭdĕ'scĕns

GROUP: *Echinocactus*.

synonyms: *Echinocactus viridescens*. Nutt. *Echinocactus limitus*. Engelm.

habitat: San Diego county, in California and southwards into the Baja California region of Mexico. Sandy and rocky hillsides, near the coast.

description: Usually a flattened globular plant, occasionally short-cylindric, up to 15 in. (37·5 cm) in diameter, and having up to 20 ribs. 4 central spines per areole in a cross formation, the lowest of which is curved downwards and up to 1½ in. (3·8 cm) in length, reddish-brown, but becoming grey in age. Radial spines mostly paler in colour, just over ½ in. (1·3 cm) long, and flattened against the body of the plant. Flower yellow or greenish-yellow usually, 1½ in. (3·8 cm) in diameter. Fruit just under 1 in. (2·5 cm) in length, red or yellow when ripe.

common name: Small Barrel Cactus.

Note. In some forms the spines change to a grey colour very quickly, as you can see in Plate 118.

plates: Nos. 118 to 120.

Ferocactus wislizenii. B. & R.

fĕ′rōkăktŭs wĭs′lĭzĕnīē

group: *Echinocactus*.

synonyms: *Echinocactus wislizenii.* Engelm.
Echinocactus emoryi. Engelm.
Ferocactus emoryi. Bckbg.

habitat: Cochise, Graham, Maricopa, Pima, Pinal, Yavapai and Yuma counties in Arizona, also in neighbouring New Mexico, El Paso county in Texas, and over the border into Sinaloa and Chihuahua in Mexico. In sandy desert, grassland and rocky areas below 6,000 ft (2,000 m).

description: A cylindrical cactus

sometimes reaching 10 ft (3·3 m) in height and up to 2 ft (60 cm) in diameter. Up to 24 or more ribs, once it has passed the juvenile stage. Up to 4 central spines, of which one is flattened and hooked occasionally exceeding 2 in. (5 cm) in length. Radial spines very variable, from very few to nearly 20 and up to 1½ in. (3·8 cm) in length and of a grey colour as are the central spines. Some of the radials rather threadlike in appearance. Flower in shades of orange, up to 2½ in. (6 cm) in diameter. Fruit is quite a feature of this species, nearly 2 in. (5 cm) in length, and yellow when ripe.

common name: Candy or Fish-hook Barrel.

plates: Nos. 80 to 82.

Hamatocactus hamatacanthus.

Bckbg & Knuth.
hămă′tŏkă′ktŭs hă′mătăkăn′thŭs

group: *Hamatocactus*.

synonyms: *Echinocactus hamatacanthus.* Mlnpfdt.
Echinocactus longihamatus. Galeotti.
Echinocactus hamatacanthus var. *longihamatus.* Coulter.
Ferocactus hamatacanthus. B. & R.

habitat: Brewster, Cameron, Duval, El Paso, Jeff Davis, Kinney, Maverick, Presidio, Upton, Val Verde, Webb and Zapata counties in Texas, and just over the border in N.E. Mexico. Desert and desert-scrub areas below 4,000 ft (1,300 m).

description: Globular to short-cylindric, solitary, up to 2 ft (60 cm) in height and 1 ft (30 cm) in diameter. Usually 13 ribs, occasionally more, bearing areoles over 1 in. (2·5 cm) apart, oval and up to ½ in. (1·3 cm) long. Usually 4 central spines, the lowest one being the longest, hooked, rounded or slightly flattened, up to 4 in. (10 cm) in length, and varying from pale brown to brownish-black occasionally. Radials between 10 and 20, up to 2 in. (5 cm) long, most of them spreading, varying in colour from straw to grey. Flower 3 in. (7·5 cm) in diameter, up to 4 in. (10 cm) high, outer petals greenish, with reddish median stripes, inner petals yellow or greenish-yellow. Fruit oblong about 1½ in. (3·8 cm) long.

COMMON NAME: Giant Fish-hook Cactus, Biznaga de Tuna, and many other names.

PLATE: No. 48.

Note: Globular, reddish-spined plant in the centre.

Hamatocactus setispinus. Engelm.
hămă′tŏkă′ktŭs sĕ′tĭspē′nŭs
GROUP: *Echinocactus.*
SYNONYMS: *Echinocactus setispinus.* Engelm.
Ferocactus setispinus. Benson.
Hamatocactus setispinus. var. *hamatus.* Engelm.
Hamatocactus setispinus. var. *setaceus.* Engelm.

Note. This is a very variable species distributed over a wide area; a number of variety names have been introduced such as the var. *setaceus* for the tall columnar form found locally in Starr county, Texas.

HABITAT: Found in some 24 counties in the south and west of Texas, and southwards across the Rio Grande, into nearby Mexico.

DESCRIPTION: Globular or cylindrical, solitary or clustering, up to 12 in. (30 cm) high and 3 in. (7·5 cm) in dia-

meter. Ribbed structure, usually 13, green body, areoles elongated. 1 central spine, standing out from the plant, hooked, and up to 1½ in. (3·8 cm) long. Radials 10–15 per areole, between ½ in. and 1 in. (1·3 cm and 2·5 cm) long, straight, spreading or flattened against the plant. All the spines tend to change from yellowish to grey in age. Flower up to 2½ in. (6 cm) in diameter, opening wide, varying shades of yellow, with a red centre. Fruit globular, ½ in. (1·3 cm) in diameter, red.

COMMON NAME: Fish-hook Cactus and other names.

PLATE: No. 139.

Note: Globular plant on right of bowl.

Hamatocactus uncinatus.
hămă′tŏkă′ktŭs ŭ′nsĭnā′htŭs
Galeotti.
GROUP: *Echinocactus.*
SYNONYMS: *Echinocactus uncinatus.* Galeotti.
Echinocactus wrightii. Engelm.
Ferocactus uncinatus. B. & R.
Echinomastus uncinatus. Knuth.
Hamatocanthus uncinatus. Galeotti.
Hamatocactus wrightii. Orcutt.
Glandulicactus uncinatus. Bckbg.
Glandulicactus uncinatus var. *wrightii.* Engelm.
Thelocactus uncinatus. Marshall.
Thelocactus uncinatus. var. *wrightii.* Kelsey & Dayton.
Ancistrocactus uncinatus. Benson.
Ancistrocactus uncinatus. var. *wrightii.* Benson.

HABITAT: Brewster, Crockett, El Paso, Jeff Davis, Presidio, Starr, Terrell and Val Verde counties in Texas, also in the Chihuahua region of Mexico.

Limestone desert and desert-scrub areas below 4,000 ft (1,300 m).

DESCRIPTION: Short columnar plant up to 8 in. (20 cm) high and about 4 in. (10 cm) diameter, usually remaining solitary. Up to 13 ribs. The body of the plant can vary from blue green to reddish brown, according to the amount of sun a plant has had to endure. Areoles oval, about ½ in. (1·3 cm) long, with a narrow groove above. Spines strong, 1 central spine standing erect, hooked at the tip and sometimes twisted, 4 in. (4 cm), occasionally up to 6 in. (15 cm) long. Up to 8 radials, spreading, some of them flattened, the longest up to 2 in. (5 cm). Flower over 1 in. (2·5 cm) in length, rarely opening wide, maroon, reddish-brown colour. Fruit oval, at least 1 in. (2·5 cm) long, red when ripe, with a few scales on it.

COMMON NAME: Cat-Claw Cactus, Brown-flowered Hedgehog and other names.

PLATES: Nos. 20 to 22.

Homalocephala texensis. B. & R.

hŏ'mălōsĕfā'lă tĕxĕ'nsĭs

GROUP: *Echinocactus.*

SYNONYMS: *Echinocactus texensis.* Hoppfer.

Echinocactus texensis fa. *longispinus.* Schelle.

HABITAT: Over a very wide area of Texas, in fact in some 25 counties, from the coast near Corpus Christi in the east, through to El Paso in the west, into southern New Mexico, also over a wide area of N.E. Mexico. Desert-scrub and grassland rarely above 3,000 ft (1,000 m).

DESCRIPTION: Solitary, flattened, globular plant, up to 12 in. (30 cm) in diameter, but even a large plant varies from only 3 in. (7·5 cm) to 8 in. (20 cm) in height. Body of plant dull green, divided up into usually around 14 ribs, but as many as 27 have been seen. Areoles about 1 in. (2·5 cm) apart, sometimes no central spines, but up to 4 are known. When present these centrals are reddish-brown to grey, straight or slightly curved, the longest of which can exceed 2 in. (5 cm) in length, and stiff with a swollen base. The radials can number up to 10 or so, but usually fewer than this, 1 in.–2 in. (2·5 cm–5 cm) in length, also very stiff, and flattened against the body of the plant, and of a similar colour to the centrals. Flower up to 2 in. (5 cm) or more in diameter, varying in colour from pale pink to shades of orange, but always with a much darker centre. Usually the edges of the petals have feathered or fringed edges. Fruit oval, exceeding 1 in. (2·5 cm) in diameter, bright red, plants often bearing six or more of them at one time.

COMMON NAME: Horse Crippler, Candy Cactus and many other names.

PLATES: Nos. 2 and 3.

Lemaireocereus thurberi. B. & R.

lĕmā'rĕōsē'rĕŭs thū'rbĕrē

GROUP: *Cereus.*

SYNONYMS: *Cereus thurberi.* Engelm.

Pilocereus thurberi. Rumpler.

Marshallocereus thurberi. Bckbg.

HABITAT: Pima and Pinal counties in Arizona, also in Baja California and western Sonora regions of Mexico. Stony desert and rocky hillsides mainly below 3,000 ft (1,000 m).

DESCRIPTION: Large, free branching, columnar cactus, up to about 20 ft (6·6 m) in height. 12–20 ribs on a stem, but quite variable even on the same specimen, with brown felted areoles. Stems up to 8 in. (20 cm) in diameter. Spines spreading, not distinguishable between radials and centrals, 12–20, brown to black and about ½ in. (1·3 cm) long. Flower white, opening at night 2½ in. (6 cm) in diameter. Fruit round 3 in. (7·5 cm) in diameter, red.

COMMON NAME: Organ-pipe Cactus.
PLATES: Nos. 62 to 64.

Lophocereus schottii. B. & R.

lŏfŏsē'rĭŭs shŏ'ttĭē
GROUP: *Pilocereus*.
SYNONYMS: *Cereus schottii*. Engelm.
Pilocereus schottii. Lemaire.

HABITAT: Pima county, Arizona, and southwards into the Baja California and Sonora regions of Mexico.
DESCRIPTION: Large clustering plant, with columnar branches, up to 21 ft (7 m) in height and 5 in. (12·5 cm) in diameter, with usually about 7 ribs, but this can vary either way. Stems or branches green or grey green in colour. Up to 7 spreading spines, black, changing to grey, but towards the upper ends of the branches, where flowers appear, the spination is quite different. Areoles are much larger there, full of grey felt-like hairs, together with up to 40 or 50 bristle-like spines, up to 2 in. (5 cm) or more long, pinkish-grey in colour. Flower appears at night, up to 1½ in. (3·8 cm) in diameter, pale pink. More than one flower can appear from each areole. Fruit egg-shaped, over 1 in. (2·5 cm) long, red, with a few scales, hairs, etc. on the surface.

COMMON NAME: Senita Cereus or Old One.
PLATES: Nos. 126 and 127.

Lophophora williamsii. Coult.

lŏfŏfŏ'ră wĭ'llĭămsĭē
GROUP: *Echinocactus*.
SYNONYMS: *Echinocactus williamsii*. Lem.
Anhalonium williamsii. Lem.
Anhalonium lewinii. Lem.
Mammillaria williamsii. Coult.
Ariocarpus williamsii. Voss.
Lophophora lewinii. Rusby.
Lophophora texana. Fric.

HABITAT: Brewster, Jim Hogg, Presidio, Starr and Webb counties, Texas, also over quite a wide area of Mexico. Limestone desert and desert-scrub, below 3,000 ft (1,000 m) usually.
DESCRIPTION: Dwarf, solitary or clustering plant, with a deep tap-root. Body of the plant rather soft, blue green to a chalky blue colour. Varies from being tubercled to a ribbed appearance, if the latter the ribs vary from 5–13. Areoles varying from ½ in.–1 in. (1·3 cm–2·5 cm) apart, containing white wool. No spines. Flower pink, off-white, occasionally creamy-yellow, just under 1 in. (2·5 cm) in diameter. Fruit small, slim, ½ in. (1·3 cm) long, pink when ripe. One is visible above a pebble in the centre, two double-headed specimens on the left and the right of Plate 9.

COMMON NAME: Peyote, Mescal Button, Dry Whisky and other names.
PLATES: Nos. 9 and 11.

Mammillaria dioica. Brand.

mă'mĭlā'rĭă dĭō'ĭcă
GROUP: *Mammillaria*.
SYNONYMS: *Chilita dioica*. Buxbaum.
Ebnerella dioica. Buxbaum.

HABITAT: Imperial, Riverside and San Diego counties in California, and southwards into Baja California region of Mexico. Rocky hillsides near coast amongst bushes and up to 5,000 ft (1,600 m).

DESCRIPTION: Globular to short-cylindric plant, tubercled, clumping only in age. Up to 6 in. (15 cm) high and about 2 in. (5 cm) in diameter. Up to 4 dark brown central spines, of which one is hooked and about ½ in. (1·3 cm) long. Up to 18 short, light brown, radial spines, flattened against the plant. Flower about 1 in. (2·5 cm) in diameter, cream with a brownish median stripe. Fruit cylindrical 1 in. (2·5 cm) long.

COMMON NAME: Not known.

PLATE: No. 116.

Note: Small globular plant on Plate 116 below the leafy flowering *Dudleya.*

Mammillaria heyderi. Muehl. var. hemisphaerica. Engelm.

mă′mĭlä′rĭă hä′dĕrē hĕmĭsfĕ′rĭkă

GROUP: *Mammillaria.*

SYNONYMS: *Mammillaria gummifera var. hemisphaerica.* Benson.
Mammillaria heyderi. Engelm.

HABITAT: Cameron, Hidalgo, Jim Wells, Kleberg, Nueces and Starr counties in Texas, also in the Tamaulipas and Nuevo Leon regions of Mexico. Surprisingly enough, it also reappears in two counties (Jackson and Gree) in Oklahoma, completely separated from its main locality in the S.E. part of Texas. Grassland and scrub areas at low altitude.

DESCRIPTION: Flat topped, globular plant, usually solitary, tubercled appearance, up to 6 in. (15 cm) in diameter. 1 central spine about ½ in. (1·3 cm) long, often much less and up to 14 radials, spreading fractionally shorter than the central, all of which are cream with a brown tip. Flower

cream sometimes tinged with pink, with a darker median stripe on the petals. Fruit 1 in. (2·5 cm) long, red when ripe.

COMMON NAME: Little Chilis.

NOT ILLUSTRATED.

Note: Similar to *M. heyderi* var. *macdougalii.*

Mammillaria heyderi. Muehl. var. macdougalii. Benson.

mă′mĭlä′rĭă hä′dĕrē măcdōō′găliē

GROUP: *Mammillaria.*

SYNONYM: *Mammillaria macdougalii.* B. & R.

HABITAT: Cochise, Pima, Pinal and Santa Cruz counties in Arizona. Grassland and among trees between 3,000 ft and 5,000 ft (1,000 m and 1,600 m).

DESCRIPTION: Flat topped, globular plant, green in colour occasionally exceeding 6 in. (15 cm) in diameter, tubercled appearance. 1–2 short yellowish central spines, and up to 12 shorter radial spines which are off-white in colour. Flower about 1 in. (2·5 cm) in diameter, cream in colour. Fruit oblong, often over 1 in. (2·5 cm) long, yellowish-green.

COMMON NAME: Cream Cactus.

PLATES: Nos. 60 and 61.

Mammillaria lasiacantha. Engelm.

mă′mĭlä′rĭă lä′zĭăkăn′thă

GROUP: *Mammillaria.*

SYNONYMS: *Mammillaria lasiacantha var. minor.* Engelm.
Mammillaria lasiacantha var. *denudata.* Engelm.
Chilita lasiacantha. Orcutt.
Mammillaria denudata. Engelm.

HABITAT: Brewster, El Paso, Hudspeth, Pecos, Presidio and Val Verde counties in Texas, S.E. parts of New Mexico, also in neighbouring parts of Mexico. Limestone desert areas below 4,000 ft (1,300 m).

DESCRIPTION: Very dwarf plant about 1 in. (2·5 cm) in diameter, sometimes larger, with a flattened top surface, which in habitat grows almost level with the surrounding soil. Solitary usually, occasionally 2 and 3 headed plants occur. Plant is densely tubercled, and the body is almost invisible because of the dense white spination. Up to 60 or more fine spines, around ⅛ in. (3 mm in length), flattened against the plant. Because of the number of spines in such a confined area from each aerole, they are formed in a number of layers. The spines when examined under a magnifying glass, have either smooth or rough surfaces. Flowers are about ½ in. (1·3 cm) in diameter, the petals being rounded at the ends usually, white with a reddish brown median stripe on each one. Fruit clubshaped, just over ½ in. (1·3 cm) long.

COMMON NAME: Lace-spine Cactus.

Note. Some authorities believe that *M. lasiacantha* var. *denudata*. Engelm is distinct from *M. lasiacantha*. Engelm. We believe it is another very variable species, where differing forms often grow alongside one another, and are best considered as one species.

NOT ILLUSTRATED.

Mammillaria meiacantha.
Engelm.
mă′mĭlā′rĭă mā′ăkăn′thă
GROUP: *Mammillaria.*
SYNONYMS: *Mammillaria gummifera* var. *meiacantha*. Benson.
Mammillaria melanocentra var. *meiacantha*. Craig.

HABITAT: Brewster, El Paso, Jeff Davis, and Presidio counties in Texas, throughout the southern counties of New Mexico, just into Cochise county, Arizona, and in adjoining parts of Mexico. Limestone desert and grassland below 5,000 ft (1,600 m).

DESCRIPTION: Flat topped globular plant, usually solitary, up to 12 in. (30 cm) in diameter, but plants smaller than this are much more common. Dark blue green to green body with rather angular tubercles, rather similar to a pyramid. Usually 1 central spine, yellowish grey with a black tip, under ½ in. (1·3 cm) long. Radials are paler than the central spine, 6–9 in number, spreading, about ⅓ in. (9 mm) long. Flower pink, often with a darker median stripe on each petal, sometimes over 1 in. (2·5 cm) in diameter. Fruit club-shaped about 1 in. (2·5 cm) long, dark pink-red.

COMMON NAME: Little Chilis.

NOT ILLUSTRATED.

Note: Similar to *M. heyderi* var, *macdougalii*.

Mammillaria microcarpa.
Engelm.
mă′mĭlā′rĭă mĭ′crōkăr′pă
GROUP: *Mammillaria.*
SYNONYMS: *Mammillaria grahamii.* Engelm.
Chilita grahamii. Orcutt.

HABITAT: Cochise, Graham, Maricopa, Mohave, Pima, Pinal, Santa Cruz and Yavapai counties in Arizona, also in San Bernardino county, in California and down into Mexico, mostly where it borders with Arizona. Desert and grassland below 5,000 ft (1,600 m).

DESCRIPTION: Globular to short cylindrical cactus, sometimes solitary, but also clustering from near the base usually. Stems sometimes exceed 2 in. (5 cm) in diameter, and occasionally exceed 6 in. (15 cm) in height. Up to 3 central spines, but normally only 1 hooked and less than 1 in. (2·5 cm) in length and of a dark reddish-brown colour. Radials sometimes exceed 24 in number, very short, and normally white in colour. Some forms have brown radials. Flower varies from pink all over to others, such as the one shown in Plate 94, with a distinct white edge to the petals. Fruit scarlet when ripe and about 1 in. (2·5 cm) in length.

COMMON NAME: Fish-hook Cactus or Lizard Catcher.

PLATES: Nos. 79, 92 and 94.

Mammillaria oliviae. Orcutt.
mămĭlā'rĭă ŏlĭ'vĭē
GROUP: *Mammillaria.*
SYNONYMS: *Mammillaria grahamii* var. *oliviae.* Benson.
Chilita oliviae. Buxbaum.

HABITAT: Cochise, Pima and Santa Cruz counties in Arizona, and across into Mexico, in the Sonora region near to the localities in Arizona. Grows in gravelly or rocky areas and even in grassland below 3,000 ft (1,000 m).

DESCRIPTION: Globular to short cylindrical, not ribbed but densely tubercled, and due to the dense spination the plant body is almost invisible. 1–3 central spines up to ⅓ in. (8 mm) long, the lowest erect, fairly rigid, white, sometimes with a dark tip. Occasionally a few hooked central spines do appear on some specimens, but this is not normal. Up to 30 or more radial spines, white or off-white, somewhat flattened against the plant; all of them much shorter than any of the centrals. Flower up to 1 in. (2·5 cm) or more in diameter, magenta-pink, but often with a white tip. Fruit rather club-shaped, 1 in. (2·5 cm) long, red when ripe.

COMMON NAME: None.

NOT ILLUSTRATED.

Mammillaria pottsii. Scheer.
mă'mĭlā'rĭă pŏ'tsĭē
GROUP: *Mammillaria.*
SYNONYMS: *Mammillaria leona.* Poselger.
Chilita pottsii. Orcutt.
Coryphantha pottsii. Berger.
Leptocladia leona. Buxbaum.

HABITAT: Brewster and Presidio counties in Texas, also over quite a wide area of northern Mexico. Limestone desert below 3,000 ft (1,000 m).

DESCRIPTION: Cylindrical plant, solitary, but more often in clusters, usually up to 4 in. (10 cm) high and up to 1½ in. (3·8 cm) in diameter, but we have found specimens exceeding 6 in. (15 cm) in length. Fairly densely tubercled, but rather obscured by the dense spination. Central spines are divided into two groups, upper and lower. The upper ones have a brownish purple tip, curved upwards, almost ½ in. (1·3 cm) long, whilst the lower ones are straight. In all they usually number up to 12. The radials average around 35 in number, thinner than the centrals, and about one-third their length, very pale brown in colour. Some plants have spines which are white or grey at the base.

COMMON NAME: Rat-tail cactus. This name is normally used for a totally different cactus from Mexico, which has long trailing stems and large magenta-pink flowers (*Aporocactus flagelliformis.*)

Note. Recently a plant has been discovered in northern Mexico, which appears to be just a miniature form of *M. pottsii*, about 2 in. (5 cm) high or less, with very slim stems, but similar spination.

PLATE: No. 42.

Mammillaria tetrancistra.
Engelm.
mă′mĭlă′rĭă tĕ′trănsĭ′stră
GROUP: *Mammillaria.*
SYNONYMS: *Mammillaria phellosperma.* Engelm.
Phellosperma tetrancistra. Fosberg.

HABITAT: Cochise, Gila, Mohave, Pima, Pinal, and Santa Cruz counties in Arizona, Imperial, Inyo, Riverside, San Bernardino and San Diego counties in California, also in south-western Nevada and just into Utah. Sandy desert areas rarely above 2,000 ft (600 m).
DESCRIPTION: Short cylindric, solitary plant up to 6 in. (15 cm) in height, and around 2 in. (5 cm) in diameter. Densely spined, almost obscuring the tubercled body. Up to 4 hooked central spines, often only 1, over ½ in. (1·3 cm) in length, off-white at the base changing to reddish-brown towards the hooked tip. Up to 30 or more white radial spines, about ½ in. (1·3 cm) or less, in length. Flower in varying shades of pink, usually with a darker median stripe, 1 in. (2·5 cm) or more diameter. Fruit cylindrical 1 in. (2·5 cm) long, red.
COMMON NAME: Corky-Seed Pincushion.
PLATE: No. 112.

Mammillaria thornberi. Orcutt.
mă′mĭlă′rĭă thō′rnbĕrē
GROUP: *Mammillaria.*
SYNONYMS: *Mammillaria fasciculata.* Engelm.
Chilita thornberi. Orcutt.

HABITAT: Pima and Pinal counties in Arizona, also in the northern Sonoran part of Mexico. Sandy desert-scrub areas rarely above 2,000 ft (600 m).
DESCRIPTION: Very freely clumping species, up to 100 heads which rarely exceed 1 in. (2·5 cm) in diameter. Each stem is cylindrical but quite narrow at the base. Usually only 1 hooked central spine of a reddish-brown colour, but up to 20 short white radial spines. Flower pale pink with a dark median stripe on each petal. Fruit oval, just over ½ in. (1·3 cm) long, red.
COMMON NAME: Clustered Pincushion.
PLATES: Nos. 84 and 85.

Mammillaria wrightii. Engelm.
var. **wilcoxii.** Marshall.
mă′mĭlārĭă rĭ′tĭē wĭlkŏ′ksĭē
GROUP: *Mammillaria.*
SYNONYMS: *Mammillaria wilcoxii.* B. & R.
Chilita wilcoxii. Orcutt.

HABITAT: Cochise, Gila, Mohave, Pima, Pinal and Santa Cruz counties in Arizona, also in certain counties in New Mexico and El Paso county in Texas, for *M. wrightii* in its various forms. Grassland, desert scrub, and among trees up to 8,000 ft (2,600 m).
DESCRIPTION: A small globular species, up to about 3 in. (7·5 cm) in diameter, densely covered with spines. These are divided up between 1–3 erect, but not stiff, hooked central spines, brown and under 1 in. (2·5 cm) in length.

The shorter radial spines are white, and pressed against the body of the plant. They vary between 12 and 20 in number. Flower pale pink usually and just over 1 in. (2·5 cm) in diameter. Fruit round, ½ in. (1·3 cm) in diameter, pink.

COMMON NAME: Wilcox's Pincushion. *Note.* In addition to *Mammillaria wrightii* and this variety, there is also another—var. *viridiflora*. There is such considerable variation in the number of spines and colour that it would not be incorrect to call them all forms of *Mammillaria wrightii.* The flower colour could range from pale green to pink or purple.

PLATE: No. 102.

Neobesseya similis. B. & R.
nēŏbĕ′sĭă sĭ′mĭlĭs
GROUP: *Mammillaria.*
SYNONYMS: *Neobesseya missouriensis.* B. & R.
Neobesseya nuttallii. Borg.
Mammillaria similis. Engelm.
Mammillaria similis var. *caespitosa.* Engelm.
Mammillaria missouriensis var. *nuttallii.* Engelm.
Coryphantha similis. B. & R.
Coryphantha nuttallii. Rumpler.
Note. There are in fact numerous other synonyms which could be listed here, depending on whether this very variable species, which has a widespread distribution, is split up. The larger growing and freely clumping form is found in Texas.

HABITAT: From the Rocky Mountains through Colorado, Kansas, Oklahoma, South Dakota, central Texas, Louisiana, in grassland and woodland at various altitudes. The smaller form, which is often solitary, will stand severe winter weather if in a well-drained position.

DESCRIPTION: Solitary or clustering, almost globular, heads flattened on top, varying from 2 in.–4 in. (5–10 cm) in diameter, prominently tubercled, dull green in colour. Centrals not always present, but if so usually 1, hard to differentiate from the radials which vary from 10–20, spreading, grey or off-white. Flower 1 in.–2 in. (2·5–5 cm) in diameter, made up of narrow petals, varying in colour from dull green, or greenish-brown, to a rose colour, but in all forms with a darker median stripe on each petal. Petals are often fringed along their edges. Fruit almost globular, red.

COMMON NAME: Nipple Cactus.

PLATE: No. 128.

Neolloydia conoidea. B. & R.
nēōloi′dĭă cŏnoi′dĭă
GROUP: *Mammillaria.*
SYNONYMS: *Mammillaria conoidea.* DC
Echinocactus conoideus. Poselger.
Neolloydia texensis. B. & R.

HABITAT: Brewster, Culberson, El Paso, Terrell and Val Verde counties in Texas, also over a wide area of Mexico, southwards from these states in the USA. Limestone desert-scrub area below 4,000 ft (1,300 m).

DESCRIPTION: Solitary, small clusters, occasionally very large clumps. Stems cylindrical up to 4 in. (10 cm) high and up to 2 in. (5 cm) in diameter, blue-green body. Up to 12 ribs, but these are not always very clear, so that it could almost be described as having tubercled appearance, rather than ribbed. Areoles circular, small with white wool. Usually 4 central spines, one of which is down pointing, about 1 in. (2·5 cm) long, black. Radials up to 20, occasionally more, spreading ⅓ in. (9 mm) long, white. Flower, 2 in. (5 cm) in diameter, opening wide, magenta-pink. Fruit small and round yellowish becoming brown and dry when ripe.

COMMON NAME: Not known.

PLATES: Nos. 23 to 25.

Neoevansia diguetii. Marshall.

nēōvă′nsĭă dĭgĕ′tĭē

GROUP: *Cereus*.

SYNONYMS: *Cereus striatus*. Brandegee.
Wilcoxia striata. B. & R.
Cereus diguetii. Weber.
Wilcoxia diguetii. Diguet & Gllmn.
Peniocereus diguetii. Bckbg.

HABITAT: Pima county, Arizona and southwards into the Baja California and Sonora regions of Mexico, in sandy desert areas, among bushes, below 1,500 ft (500 m).

DESCRIPTION: The aerial part of the plant is very inconspicuous, with slender, usually 9 ribbed stems, up to 2½ in. (6 cm) high and ¼ in. (6 mm) in diameter. These are grey-green in colour. Up to 10 very small thin spines, flattened against the stem, off-white to pale brown in colour. Flower up to 3 in. (7·5 cm) in diameter, nocturnal usually, occasionally staying open for part of the next day, white to pink in colour. Fruit egg-shaped, up to 2 in. (5 cm) long, reddish-purple. The root of this plant consists of a number of swollen tuberous roots similar to those of a dahlia, and like most *Wilcoxias*, see diagram, p. 230.

COMMON NAME: Dahlia-rooted Cereus.

Note. There is considerable disagreement over the correct genus for this plant, as it seems to fall midway between *Peniocereus* and *Wilcoxia*.

NOT ILLUSTRATED.

Opuntia acanthocarpa. Engelm. & Bigelow.

var. **thornberi.** Benson.

ōpŏŏ′ntĭă ăkănthōkā′rpă thŏ′rnbĕrē

GROUP: *Opuntia*.

SYNONYMS: *Opuntia thornberi*. Thornber & Bonker.
Cylindropuntia acanthocarpa var. *thornberi*. Bckbg.

HABITAT: Gila, Graham, Maricopa, Mohave, Pinal and Yavapai counties in Arizona. Rocky hillsides around 3,000 ft (1,000 m).

DESCRIPTION: Low-growing and free branching shrub, which occasionally attains 6 ft (2 m) in height. Joints up to 20 in. (50 cm) in length, but often much less, and usually about 1½ in. (3·8 cm) in diameter. Up to 12 or more 1 in. (2·5 cm) long spines, straw-yellow in colour, but as areoles are well apart, the green to pinkish-brown body colour shows through. Flower very variable in colour from yellow to red, and 2 in. (5 cm) in diameter. Fruit oval, up to 1½ in. (3·8 cm) long, tuberculate, brown.

COMMON NAME: Thornber Cholla.

NOT ILLUSTRATED.

Opuntia arbuscula. Engelm.

ōpŏŏ′ntĭă ărbŏŏ′scŭlă

GROUP: *Opuntia*.

SYNONYMS: *Opuntia vivipara*. Rose.
Cylindropuntia arbuscula. Knuth.

HABITAT: Cochise, Maricopa, Pima, Pinal, Santa Cruz, Yavapai and

Yuma counties in Arizona, and in Mexico in the Sinaloa and Sonora regions, in sandy flat desert areas, usually below 3,000 ft (1,000 m).

DESCRIPTION: Low-growing and free branching shrub, up to 5 ft (1·5 m) in height. Joints up to 6 in. (15 cm) long, often less, and less than ½ in. (1·25 cm) across. Joint surface rather smooth, and only slightly raised at the areole positions. Up to 4 spines per areole, often much less, up to 1½ in. (4 cm) long, straw coloured, longest one pointing downwards. Joints vary in colour from greenish-pink to purplish (sometimes). Flower about 1 in. (2·5 cm) across, very variable in colour from greenish-yellow through to a brick-red. Fruit rather pear-shaped, without spines, green.

COMMON NAME: Pencil Cholla.

NOT ILLUSTRATED.

Opuntia basilaris. Engelm. & Bigelow.

ōpŏŏ′ntĭă băsĭlār′ĭs

GROUP: *Opuntia.*

SYNONYM: *Opuntia intricata.* Griffith.

Opuntia basilaris var. *albiflorus.* Walton.

HABITAT: Coconino, Maricopa, Mohave, Yavapai and Yuma counties in Arizona, also in southern Nevada and Utah, Kern, Imperial, Inyo, Los Angeles, Orange, Riverside, San Bernardino, San Diego and Ventura counties in California, and southwards into the Sonora region of Mexico. From sandy desert to hillside grassland areas between sea level and 9,000 ft (3,000 m).

DESCRIPTION: Low growing clumps, somewhat prostrate habit, as new pads are able to root down into the sandy soil. These pads are oval shaped

up to 12 in. (30 cm) in length, 4 in. (10 cm) or more wide, and at the most ½ in. (1·3 cm) thick, blue-grey colour. There are no spines, only areoles full of very short brown glochids. Flower up to 3 in. (7·5 cm) in diameter of a mauvy-pink colour. Fruit oval, just over 1 in. (2·5 cm) long, grey, with a few glochids.

COMMON NAME: Beaver Tail Cactus or Prickly Pear.

Note. There are three varieties of *O. basilaris* which vary in pad size, and shape and have differing flower colours.

PLATE: No. 106.

Opuntia bigelovii. Engelm.

ōpŏŏ′ntĭă bĭ′gĕlŏ′vĭē

GROUP: *Opuntia.*

SYNONYM: *Cylindropuntia bigelovii.* Knuth.

HABITAT: Gila, Maricopa, Mohave, Pima, Pinal, Yavapai and Yuma counties in Arizona, also in the northern Baja California and Sonora regions of Mexico. Desert to rocky hillsides up to 3,000 ft (1,000 m).

DESCRIPTION: Low-growing tree-like plants, sometimes exceeding 6 ft (2 m) in height, with most of the branches on the upper half. Old part of the stem becomes black in appearance because the spines change to this dirty colour from their original silvery-golden colour. Densely spined; spines sometimes exceed 8 per areole, and are about 1 in. (2·5 cm) long and barbed. Joints average around 6 in. (15 cm) in length, and up to 2 in. (5 cm) usually in diameter. Flower 1½ in. (3·8 cm) in diameter, variable colour from off white to pale green or even pale lavender. Fruit oval, tubercu-

late, up to 1 in. (2·5 cm) long, yellowish-green.
COMMON NAME: Teddy Bear Cholla.
PLATES: Nos. 89 to 91.

Opuntia chlorotica. Engelm & Bigelow.
ōpŏŏ'ntĭă klŏrŏ'tĭkă
GROUP: *Opuntia*.
SYNONYM: None.

HABITAT: All the counties of Arizona, except Apache and Navajo, also in Riverside, San Bernardino and San Diego counties in California, and southwards into the Baja California and Sonora regions of Mexico. Rocky desert to woodland up to 6,000 ft (2,000 m).
DESCRIPTION: Somewhat tree-like in habit, but rarely exceeding 8 ft (2·6 m) in height, with a definite tree-like trunk, which becomes cylindrical in age. The joints or pads are almost circular and up to 8 in. (20 cm) in diameter, and blue green in colour. Up to 6 yellow spines, 1½ in. (3·8 cm) in length, pointing downwards. The areoles are armed with many yellow glochids. Flower often exceeds 2 in. (5 cm) in diameter, pale yellow, and with a slight brownish median stripe on each petal, but this is not always present. Fruit oval, just under 1 in. (2·5 cm) long, tuberculate, yellow or yellowish-green.
COMMON NAME: Clock-face Prickly Pear.
PLATE: No. 105.

Opuntia engelmannii. Salm-Dyck.
ōpŏŏ'ntĭă ĕn'gĕlmă'nĭē
GROUP: *Opuntia*.
SYNONYMS: *Opuntia engelmannii* var. *cyclodes*. Engelm.

Opuntia lindheimeri var. *cyclodes*. Counter.
Opuntia discata. Griffiths.
Opuntia phaeacantha var. *discata*. Benson & Walk.

Also numerous other names depending on whether one can accept as valid the many so-called varieties of *O. engelmannii* which are listed for this very variable and widespread species.

HABITAT: Very widespread in its many forms over Texas, New Mexico, Arizona, and in all the regions of Mexico, bordering with the USA. All types of habitats up to 5,000 ft (1,600 m).
DESCRIPTION: A wide spreading bushy plant, up to 5 ft (1·6 m) or so in height, with circular or slightly oblong pads or joints. These can reach up to 10 in. (25 cm) in length and diameter and a little over ½ in. (1·3 cm) thick. Pads vary in colour from green to blue green often with a powdery bloom on them as on the surface of a fruit such as a grape. Areoles are not close together; some have yellow or brownish glochids. Spination is also very variable in colour as with the glochids, 4 and 10, occasionally more, and ranging from 1 in. to 2 in. (2·5 cm–5 cm) in length. The direction of the spines is also very variable. Flower orange or yellow usually, red in some forms, 3 in.–4 in. (7·5 cm–10 cm) in diameter. Fruit when ripe in varying shades of red or purple, pear-shaped and up to 3 in. (7·5 cm) long.
COMMON NAME: Nopal, Tuna and Engelmann's Prickly Pear.
PLATES: Nos. 8, 31 and 32.

Note: A young two-padded plant is visible on the left of Plate 8, along with *Astrophytum asterias*.

Opuntia fulgida. Engelm.
ōpŏŏ'ntĭă fŭ'ljĭdă
GROUP: *Opuntia.*
SYNONYM: *Cylindropuntia fulgida.* Knuth

HABITAT: Cochise, Graham, Maricopa, Pima, Pinal and Yuma counties in Arizona, also in the northern part of Baja California and northern Sonora regions of Mexico. Rocky deserts and hillsides up to 3,000 ft (1,000 m).
DESCRIPTION: Tree-like cholla *Opuntia*, up to 15 ft (5 m) in height, which branches very freely. Cylindrical joints up to 6 in. (15 cm) in length, easily detached from the parent plant. Spines per areole very variable from 2–12, spreading, off-white to yellow in colour, exceeding 1 in. (2·5 cm) in length, and barbed. Flower in varying shades of lavender. Flower forms out of a fruit from a previous year, so that over a period of years a chain fruit system is formed. Fruit oval, nearly 1½ in. (3·8 cm) long, green.
COMMON NAME: Chain Fruit Cholla and Jumping Cholla.
PLATE: No. 83.

Opuntia fulgida var. **mamillata.** Coulter.
ōpŏŏ'ntĭă fŭ'ljĭdă mă'mĭlă'htă
GROUP: *Opuntia.*
SYNONYMS: *Cylindropuntia fulgida* var. *mamillata.* Bckbg.
Opuntia mamillata. Schott.

HABITAT: Pima and Pinal counties in Arizona, also in the northern part of Baja California and northern Sonora regions of Mexico, in rocky deserts and hillsides up to 3,000 ft (1,000 m).
DESCRIPTION: This variety only reaches 5 ft (1·6 m) in height, and has weaker drooping branches. Up to 6 spines per areole, of similar length to the type species, but being less densely spined, the joints tend to be pinker in colour instead of green.
COMMON NAME: Smooth Chain Fruit Cholla.
PLATES: Nos. 77 and 78.

Opuntia imbricata. DC
ōpŏŏ'ntĭă ĭm'brĭcă'htă
GROUP: *Opuntia.*
SYNONYMS: *Opuntia arborescens.* Engelm.
Opuntia vexans. Griffiths.
Cylindropuntia imbricata. Knuth.

HABITAT: Bailey, Bexar, Brewster, Burnet, Crockett, Culberson, Dickens, El Paso, Howard, Hudspeth, Jeff Davis, Mitchell, Nolan, Oldham, Pecos, Potter, Presidio, San Saba, Tom Green and Victoria counties in Texas, more northerly counties of New Mexico, Colorado, Kansas, Oklahoma, and into Cochise, Gila and Pima counties in Arizona. Also over quite a wide area of northern Mexico. Low-lying areas, desert-scrub and hillsides up to 3,000 ft (1,000 m).
DESCRIPTION: Tree-like, free branching shrub up to 9 ft (3 m) in height, but 5 ft–6 ft (1·6 m to 2 m) is more common. Joints usually green, between 8 in. and 12 in. (20 cm and 30 cm) in length, and about 1 in. (2·5 cm) in diameter. The tubercles on the joint are rather elongated, up to 2 in. (5 cm) in length. Very few glochids per areole, but up to 30 spines between ½ in. and 1 in. (1·3 cm and 2·5 cm) long, cream or brown in colour, sometimes tinged with red. Flower in varying magenta shades, opening wide, about 2 in. (5 cm) or more diameter. Fruit very tuberculate, like the

I

stems, spiny, about 1½ in. (3·8 cm)
long, yellow when ripe.
COMMON NAME: Cane Cactus, Cane
Cholla and other names.
PLATE: No. 45.

Opuntia kleiniae. DC
var. **tetracantha.** Marshall.
ŏpŏŏ'ntĭă klī'nĭē tĕtrăkă'nthă
GROUP: *Opuntia.*
SYNONYMS: *Opuntia tetracantha.* Toumey.
Cylindropuntia tetracantha. Knuth.
Opuntia californica. Engelm.

HABITAT: Cochise, Gila, Navajo,
Pima, Pinal and Yavapai counties in
Arizona and southwards into the
Sinaloa and Sonora regions of Mexico.
Desert to grassland up to 4,000 ft
(1,300 m).
DESCRIPTION: Freely branching semi-
sprawling bush up to 4 ft (1·3 m) in
height. Pinkish-red joints up to 10 in.
(25 cm) in length, and only ½ in. (1·3
cm) in diameter. Up to 6, but usually
4 pale brown spines, about 1 in. (2·5
cm) long. *Opuntia kleiniae* always has
fewer spines per areole, often only 1,
and is a more erect shrub or bush up
to 7 ft (2·3 m) in height. In the var.
tetracantha the 2 in. (5 cm) diameter
flowers are partly green, with brownish edges, instead of being purple.
Fruit oval, ¾ in. (2 cm) long, red.
COMMON NAME: Klein's Cholla.
NOT ILLUSTRATED.

Opuntia leptocaulis. DC
ŏpŏŏ'ntĭă lĕ'ptŏcă'wlĭs
GROUP: *Opuntia.*
SYNONYMS: *Opuntia frutescens.* Engelm.
Opuntia fragilis var. *frutescens.* Engelm.
Cylindropuntia leptocaulis. Knuth.
Note: A number of other varieties have
been described of *O. leptocaulis,* because

of its variability, and these also become
synonyms.

HABITAT: Over a very wide area from
Arizona, through New Mexico, and
much of Texas, also northern Mexico.
Deep soil areas and desert-scrub up to
4,500 ft (1,500 m).
DESCRIPTION: Small upright shrub, up
to 5 ft (1·6 m) in height, but usually
much less, freely branching. Cylindrical joints up to 12 in. (30 cm) long,
and barely ¼ in. (6 mm) thick.
Areoles oval shaped less than ⅓ in. (9
mm) apart. Spines, usually 1 per
areole, under 2 in. (5 cm) long, downward pointing, straw to pale brown in
colour, becoming grey in age. The
main stem or trunk of these shrubs
becomes barky, losing most of its
spines, and is well over 1 in. (2·5 cm)
in diameter. Flower barely 1 in. (2·5
cm) in diameter, greenish-yellow.
Fruit oval, with glochids, about ½ in.
(1·3 cm) long, red.
COMMON NAME: Tasajillo, Desert
Christmas Cactus and other names.
PLATE: No. 27.

Opuntia macrocentra. Engelm.
ŏpŏŏ'ntĭă mă'crōsĕntră
GROUP: *Opuntia.*
SYNONYM: *Opuntia vilacea* var. *macrocentra.* Benson.

HABITAT: Brewster, El Paso, Hudspeth, Jeff Davis, Presidio, Reeves and
Terrell counties in Texas, in some of
the southern counties of New Mexico,
Cochise and Pima counties in Arizona,
and in the northern part of the Chihuahua region of Mexico. Rocky
desert to desert-scrub up to 5,000 ft
(1,600 m).
DESCRIPTION: Upright, bushy habit,
up to 3 ft (1 m) in height, but often

only half this height, free branching, and not forming a central trunk as do many of the flat padded *Opuntias*. Pads or joints are almost circular, up to 8 in. (20 cm) in diameter, no more than ½ in. (1·3 cm) thick, blue green or tinged with red or purple. Areoles small, about ¾ in. (2 cm) apart, with a few brown glochids. Usually only 1 or 2 spines per areole, occasionally up to 4, between 2 in. and 6 in. (5 cm and 15 cm) long, usually flattened, and black. These spines are quite flexible. Flower wide opening, 3 in. (7·5 cm) in diameter, yellow with a reddish centre. Fruit oval, over 1 in. (2·5 cm) long.

COMMON NAME: Purple Prickly Pear.
PLATE: No. 27. Only visible as the stick-like stems in the background.

Opuntia oricola. Philbrick.

ōpōō′ntiă ōrī′kŏlă
GROUP: *Opuntia*.
SYNONYM: None.

HABITAT: Los Angeles, Orange, San Diego, Santa Barbara and Ventura counties in California, and on some of the offshore islands, also into northern part of the Baja California region of Mexico. Sandy areas, hillsides of the coastal Chapperal region.
DESCRIPTION: Tree-like plant, freely branching up to 10 ft (3·3 m) in height, made up of joints or pads up to 10 in. (25 cm) in length and slightly less in width and ½ in. (1·3 cm) thick. Few glochids, but up to 16, 1 in. (2·5 cm) long spines, yellow when young, changing to brown and grey later. Flower around 2 in. (5 cm) in diameter, yellow, with a fairly prominent green stigma. Prominent red fruits, just over 1 in. (2·5 cm) long.

COMMON NAME: Prickly Pear. (One of many *Opuntias* which have this common name.)
PLATE: No. 121.

Opuntia rufida. Engelm.

ōpōō′ntiă rōō′fĭdă
GROUP: *Opuntia*.
SYNONYM: *Opuntia microdasys* var. *rufida*. Schumann.

HABITAT: Brewster and Presidio counties in Texas, also in the Chihuahua and Coahuila regions of Mexico. Desert and rocky hillsides up to 3,000 ft (1,000 m).
DESCRIPTION: An erect, free branching shrub, forming a central trunk in age, but not exceeding 6 ft (2 m) in height. The joint or pad shape is quite variable over its range from oval to almost circular, but the latter shape is the type we found in Texas. These grey-green pads were 4 in. and 6 in. (10 cm and 15 cm) in diameter. Areoles, circular, about ¼ in. (6 mm) in diameter, bearing a large number of very short reddish-brown glochids. No spines. Flower, about 2 in. (5 cm) in diameter, yellow, opening wide. Fruit almost circular, 1 in. (2·5 cm) in diameter, greenish-brown, also with brown glochids.

COMMON NAME: Blind Prickly Pear.
PLATE: No. 139. Flat padded plant at rear of bowl.

Opuntia santa-rita. Rose.

ōpōō′ntiă să′tă-rē′tă
GROUP: *Opuntia*
SYNONYMS: *Opuntia chlorotica* var. *santa-rita*. Griffiths & Hare.
Opuntia gosseliniana var. *santa-rita*. L. Benson.
Opuntia shreveana. Nelson.

HABITAT: Cochise, Gila, Pima and Santa-Cruz counties in Arizona, the southern-most counties of New Mexico, and Brewster, Jeff Davis and Presidio counties in Texas and over the border into the northern Sonora region of Mexico, in desert grassland and woodland below 5,000 ft (1,600 m).

DESCRIPTION: Upright shrub with a short trunk, rarely exceeding 5 ft (1·5 m) in height, and bearing numerous fairly thin, flat and almost circular pads, which are up to 7 in. (18 cm) in diameter. They vary in colour from greenish-blue, to wonderful shades of pink or even pale violet-purple. Areoles about 1 in. (2·5 cm) apart, full of reddish-brown glochids. Spines when present can be up to 4 in. (10 cm) in length and of a similar colour. Flower about 3 in. (7·5 cm) across, pale yellow. Fruit is rather slim, 1 in. (2·5 cm) in length and purple when ripe.

COMMON NAME: Purple Prickly Pear.

PLATES: Nos. 86 and 87.

Opuntia schottii. Engelm.
ōpŏŏ'ntĭă shŏ'tĭē
GROUP: *Opuntia.*
SYNONYM: *Corynopuntia schottii.* Knuth.

HABITAT: Brewster, Cameron, Hidalgo, Kinney, Maverick, Starr, Terrell, Webb, Val Verde and Zapata counties in Texas, and in close proximity to the Rio Grande on the Mexican side of the border, between Matamoros and the Big Bend National Park. Sandy areas up to 3,000 ft (1,000 m).

DESCRIPTION: Creeping, free branching plant, often growing among grass, such that it is not easily seen. Club-shaped joints up to 3 in. (7·5 cm) long, and 1 in. (2·5 cm) in diameter at the widest point, which is near the upper end. Joints have elongated tubercles up to ¾ in. (2 cm) long. Up to 14 spines per areole, the central one of which is very strong, very flattened erect up to 2½ in. (6 cm) long, the surface of which is ridged, and pale yellow in colour, often with a white edge. There are 2 other spines similar to this one in appearance, but downward pointing and only 1½ in. (3·8 cm) long. The remainder of the spines are usually rounded, about ½ in. (1·3 cm) long and white, but some of them tend to be flattened in appearance. Few straw-coloured glochids. Flower yellow, up to 2½ in. (6 cm) in diameter, and 2 in. (5 cm) long. Fruit oval, up to 1½ in. (3·8 cm) long, spiny, yellow when ripe.

COMMON NAME: Devil Cactus or Dog Cholla.

PLATE: No. 12.

Opuntia schottii. Engelm.
var. **grahamii.** Benson.
ōpŏŏ'ntĭă shŏ'tĭē grā'ămē
GROUP: *Opuntia.*
SYNONYMS: *Corynopuntia grahamii.* Knuth.
Opuntia grahamii. Engelm.

HABITAT: Brewster, El Paso and Presidio counties in Texas, also just into the southern part of New Mexico. Sandy desert and desert-scrub up to 3,000 ft (1,000 m).

DESCRIPTION: Creeping, free branching plant, forming dense mats of club-shaped joints, about 2 in. (5 cm) long, covered in low tubercles ½ in. (1·3 cm) long. Up to 14 spines, separated into an outer and inner group, up to 2 in.

(5 cm) long. The inner ones are some-what thicker, spreading in all direc-tions and as with the others brown when young, changing to grey. The areoles contain numerous brown glochids. Flower yellow, exceeding 2 in. (5 cm) in diameter, standing up above the matt-formation of the plant. Fruit oval, 1½ in. (3·8 cm) long, spiny, yellow when ripe.
COMMON NAME: Graham Dog Cactus or Mounded Dwarf Cholla.
PLATE: No. 33.

Opuntia spinosior. Toumey.
ŏpŏŏ'ntĭă spĭnŏ'sĭŏr
GROUP: *Opuntia.*
SYNONYM: *Opuntia whipplei* var. *spino-sior.* Engelm.

HABITAT: Cochise, Gila, Graham, Maricopa, Pima, Pinal, Safford and Santa Cruz counties in Arizona, and on into nearby New Mexico, as well as the Chihuahua and Sonora regions of Mexico, on grassland usually up to 6,000 ft (3,000 m).
DESCRIPTION: Shrubby, tree-like habit up to 12 ft (4 m) in height, forming a distinct woody trunk, which can reach as much as 9 in. (23 cm) across. Joints up to 12 in. (30 cm) long, often less, about ¾ in. (2 cm) across, distinctly tubercled, but in rows forming a type of rib structure. These elongated tubercles are about ½ in. (1·3 cm) long, bearing up to 20 spines, just over ¼ in. (6 mm) long, grey to pinkish-red. The areoles themselves bear a little wool and a few glochids. Flower up to 2 in. (5 cm) across, very variable in colour from white, through shades of yellow to red or purple. Fruit egg-shaped, about 1½ in. (3·8 cm) long, yellow.

COMMON NAMES: Spiny Cholla and Cane Cholla.
NOT ILLUSTRATED.

Opuntia versicolor. Engelm.
ŏpŏŏ'ntĭă vĕ'rsĭcŏlŏr
GROUP: *Opuntia.*
SYNONYMS: *Opuntia arborescens* var. *versicolor.* E. Dams.
Cylindropuntia versicolor. E. M. Knuth.

HABITAT: Cochise, Gila, Pima and Pinal counties in Arizona, and in the nearby Sonora region of Mexico, in sandy desert areas below 3,000 ft (1,000 m).
DESCRIPTION: Small tree-like species, up to 12 ft (4 m) high, but usually much less. Freely branching, but with a small trunk at the base. Branches or joints, about 8 in. (20 cm) long and ¾ in. (2 cm) in diameter, sometimes slightly more. Joints tubercled, elon-gated and 1 in. (2·5 cm) long. Up to 10 barbed spines per areole, spreading, varying in length, but around ½ in. (1·3 cm) long. Some brown glochids present also on each areole. Flower up to 1½ in. (3·8 cm) in diameter, opening wide, varying in colour from greenish brown to red. Fruit egg-shaped, over 1 in. (2·5 cm) long, greenish-yellow when ripe, few spines, but tending to form chains of them as with *O. fulgida.*
COMMON NAME: Staghorn Cholla.
NOT ILLUSTRATED.

Pediocactus knowltonii. Benson.
pē'dĭŏkă'ktŭs nŏltŏ'nĭē
GROUP: *Echinocactus.*
SYNONYM: *Pediocactus bradyi* var. *knowl-tonii.* Bckbg.

HABITAT: La Boca, and along Los

Pinos river, Colorado and just into nearby New Mexico.

DESCRIPTION: Dwarf, globular, solitary plant, and small clusters, single heads rarely exceeding 1 in. (2·5 cm) in diameter. Tubercled, pale green body, no central spines, but around 20 very short, pectinate, white radials. Flower just over ½ in. (1·3 cm) in diameter, pale pink. Fruit oval, grey-green, dry when ripe.

COMMON NAME: Not known.

PLATE: No. 129.

Pediocactus simpsonii. B. & R.

pē'dĭŏkă'ktŭs sĭ'mpsōnĭē

GROUP: *Echinocactus.*

SYNONYMS: *Echinocactus simpsonii.* Engelm.

Mammillaria simpsonii. Jones.

Pediocactus simpsonii. var. *hermannii.* Wgnd & Bckbg.

HABITAT: Nr Grand Canyon, Arizona, northern New Mexico, Idaho, Kansas, Montana, Nevada, Oregon and Washington, and Utah mostly on hills in woodland, between 6,000 ft and 10,000 ft (2,000 m and 3,300 m).

DESCRIPTION: Globular or flattened globular plant, solitary or small clusters. Tubercled, green, up to 8, ¾ in. (1·3 cm) long central spines, reddish-brown in colour. Up to 25 radials, shorter than centrals, spreading, off-white. Flower off-white to pale pink, 1 in. (2·5 cm) in diameter. Fruit almost globular ½ in. (1·3 cm) in diameter, green.

COMMON NAME: Plains Cactus.

Note. This is quite a hardy cactus, standing very low winter temperatures if dry, but prone to rotting in cultivation if care is not taken.

PLATE: No. 130.

Peniocereus greggii. B. & R.

pĕ'nĭŏsē'rĭŭs grē'gĭē

GROUP: *Cereus.*

SYNONYMS: *Cereus greggii* var. *cismontanus.* Engelm.

Cereus greggii var. *transmontanus.* Engelm.

Cereus greggii. Engelm.

Cereus greggii var. *roseiflorus.* Kuntze.

HABITAT: Brewster, Hudspeth, Jeff Davis, Pecos, Presidio and Terrell counties in Texas, most of the southern counties of New Mexico, Cochise, Graham, Maricopa, Pima, Pinal and Yuma counties in Arizona, and in the Chihuahua and Zacatecas regions of Mexico. Sandy desert-scrub area up to 2,000 (600 m).

DESCRIPTION: The aerial part of the plant is very inconspicuous, with slender 4–5 ribbed stems up to 6 ft (2 m) long, but often much less. The number of stems or branches from one plant can vary from 1 to a dozen or so occasionally, these tending to sprawl unless supported on the branches of a nearby bush or tree, as they are no more than ½ in. (1·3 cm) in diameter. Stems grey green, blue green sometimes tinged with red, with very small areoles close together on the ribs, bearing between 6 and 12 black spines, changing to grey, but only about ⅛ in. (3 mm) long. The stems appear from a large tuberous root which can weigh many pounds (see p. 50). Flowers at night, up to 3 in. (7·5 cm) in diameter, and up to 8 in. (20 cm) long, white. The outer petals are brownish tinged, and the outer surface of the flower tubs has bristle-like spines and scales. Fruit oval, with a few short black spines, orange-red.

COMMON NAME: Arizona Queen of the Night, Deer-horn Cactus and many other names.

PLATES: Nos. 50 to 52.

Note: The specimen shown with its bare tuberous root in Plate 50 was in fact growing just to the right of the mesquite bush on the left of the picture.

Sclerocactus whipplei. B. & R.
sklē'rōkăktŭs wĭ'plīē

GROUP: *Echinocactus.*

SYNONYMS: *Echinocatus whipplei.* Engelm & Bigelow.
Echinocactus whipplei var. *spinosior.* Engelm.
Sclerocactus whipplei var. *pygmaeus.* Peebles.

HABITAT: Apache and Navajo counties in Arizona, also just over the border into Utah and Colorado, at around 5,000 ft (1,600 m) altitude, on rocky hillsides and canyons.

DESCRIPTION: Solitary, short-cylindrical plant, up to 12 in. (30 cm) high and 4 in. (10 cm) in diameter, occasionally even larger. Rib structure after the juvenile stage, often in a spiral, with variable coloured spines from white through to red. Up to 3 central spines, hooked, up to 2 in. (5 cm) long, fairly erect. Up to 11 radials, spreading, and tending to intertwine up to 1½ in. (3·8 cm) long, some of these slightly hooked or curved. Flower up to 2 in. (5 cm) in diameter, varying from plant to plant, greenish-yellow through to shades of purple. Fruit globular about ½ in. (1·3 cm) in diameter, pinkish-brown; seeds are released by a pore at the base.

COMMON NAME: Devil's Claw Barrel.

NOT ILLUSTRATED.

Thelocactus bicolor. B. & R.
var. schottii. B. & R.
thĕ'lōkăktŭs bī'cŏlŏr shŏ'tīē

GROUP: *Echinocactus.*

SYNONYMS: *Echinocactus bicolor.* var. *schottii.* Engelm.
Echinocactus schottii. Small.
Thelocactus bicolor. var. *texensis.* Bckbg.

HABITAT: Brewster and Starr counties in Texas, also in the Tamaulipas region of Mexico, in stony desert-scrub and grassland areas below 4,000 ft (1,300 m).

DESCRIPTION: Cone-shaped to short columnar plant usually about 6 in. (15 cm) high and 4 in. (10 cm) in diameter, but occasionally taller and a little wider. Solitary, but occasionally 2–3 heads, made up of 8 ribs. Areoles rather elongated, in the formation of a groove, about ¼ in. (6 mm) long. Up to 4 central spines, up to 2 in. (5 cm) or more length, one of which is flattened. Radials pressed against the plant, about 15, a few of these sometimes flattened, about 1½ in. (3·8 cm) long, and as with the centrals, yellowish when young, changing to grey. Flower often over 3 in. (7·5 cm) in diameter, shiny pink on the petals, but very variable, with a red throat. Fruit ½ in. (1·3 cm) long, brown, dry, losing its seeds by splitting at the base so that they run out.

COMMON NAME: Texas Pride or Glory of Texas.

PLATE: No. 7.

Toumeya papyracantha. B. & R.
tŏŏmā'yă pă'pīrăkăn'thă

GROUP: *Echinocactus.*

SYNONYMS: *Mammillaria papyracantha.* Engelm.
Echinocactus papyracanthus. Engelm.
Pediocactus papyracanthus. Benson.

HABITAT: Apache and Navajo counties in Arizona, and Navajo county, New Mexico, in grassland areas around 6,000 ft (2,000 m).

DESCRIPTION: Solitary, cylindrical cactus usually, up to 3 in. (7·5 cm) high, and just over ½ in. (1·3 cm) in diameter. Soft bodied, blue-green, and brownish without spines nearer the base. Tubercled, soft almost papery spined, off-white to pale brown. Up to 4 central spines, at least one of which can be over 1 in. (2·5 cm) long, and flattened. Radials paler colour, up to 8, against the body of the plant, under ¼ in. (6 mm) long. Flower up to 1 in. (2·5 cm) in diameter, appears much less as it does not open wide, white with brown or greenish-brown median stripes on the petals. Fruit nearly globular, less than ¼ in. (6 mm) in diameter, brown and dry when ripe.

COMMON NAME: Gramma Grass Cactus

PLATE: No. 131.

Utahia sileri. B. & R.

ūtā′ĭă sīlē′rē

GROUP: *Echinocactus.*

SYNONYMS: *Echinocactus sileri.* Engelm. *Pediocactus sileri.* Benson.

HABITAT: Mohave county in Arizona and just into Utah, in desert areas and on hills around 5,000 ft (1,600 m).

DESCRIPTION: Globular to short-cylindrical plant, up to 5 in. (12·5 cm) high and just under 4 in. (10 cm) in diameter. Spirally ribbed usually, only tubercled when young, blue-green to nearly black body. Up to 7 central spines, usually less, standing out from the plant, pale brown to black towards the tips, and about 1 in. (2·5 cm) long. Up to 16 radials, white, spreading, about ½ in. (1·3 cm) long. Flower about 1 in. (2·5 cm) in diameter, greenish-yellow to very pale yellow. Fruit almost globular green.

COMMON NAME: Gypsum cactus. (The soil in its habitat is rich in gypsum.)

PLATE: No. 132.

Wilcoxia poselgeri. B. & R.

wĭlkŏ′ksĭă pŏ′sĕlgĕrē

GROUP: *Cereus.*

SYNONYMS: *Cereus poselgeri.* Coulter. *Cereus tuberosus.* Poselger. *Echinocereus poselgeri.* Lem. *Echinocereus tuberosus.* Rumpl.

HABITAT: Cameron, Starr, Webb and Zapata counties in Texas, in desert-scrub regions, among bushes at low altitudes.

DESCRIPTION: One or more slender stems, with only a few branches above ground, cylindrical up to 3 ft (1 m) occasionally higher, but at the most ⅓ in. (9 mm) in diameter. Up to 10 ribs, areoles close-together, bearing 1 central spine, ¼ in. (6 mm) long, pointed upwards, white with a brown tip and up to 14 or so radials, flattened against the stem, white or grey in colour. Flower up to 2 in. (5 cm) in diameter, dark pink, opening wide. Fruit oval, under 1 in. (2·5 cm) long, covered with spines and bristles. The root system consists of a number of tuberous roots, such as dahlias have, hence its common name. See p. 230 (diagram).

COMMON NAME: Dahlia Cactus or Pencil Cactus.

PLATE: No. 13.

FAMILY: AMARYLLIDACEAE
(but often now referred to as under Agavaceae).

Agave lophantha. Schied.
var. **poselgeri.** Berger.
ăgā'hvĭ lōfă'nthă pŏ'sĕlgĕrē
GROUP: *Agave.*
SYNONYM: *Agave lechuguilla.* Berger.

HABITAT: Brewster, Crockett, Culberson, Jeff Davis, El Paso, Pecos, Presidio, Terrell and Val Verde counties in Texas, and over quite a wide area of the Chihuahuan desert region of Mexico below 5,000 ft (1,600 m).

DESCRIPTION: Stemless succulent, clustering, with the leaves held fairly erect, these up to 10 in. (25 cm) long, and about ⅓ in. (9 mm) across at the widest point. Leaves very stiff, as is also the grey thorn at the leaf tip. The leaf margin is grey, with small curved thorns at intervals, whilst the remainder of the leaf surface is olive-green with scattered grey-brown markings on them. Flower spike is up to 15 ft (5 m) in height, unbranched, bearing 2 in. (5 cm) long reddish-brown flowers. Seed pod, cylindrical, just under 1 in. (2·5 cm) in length, grey.

COMMON NAME: Lechuguilla.
PLATE: No. 46.
Note: Green-leaved plant right of centre.

Agave palmeri. Engelm.
ăgā'hvĭ pā'lmĕrē
GROUP: *Agave.*
SYNONYMS: None.

HABITAT: Cochise, Gila, Graham, Maricopa, Pima, Pinal, Santa Cruz and Yuma counties in Arizona, also in the counties in south-western New Mexico and northern Sonora region of Mexico, in many types of habitat below 5,000 ft (1,600 m).

DESCRIPTION: Stemless succulent up to 4 ft (1·3 m) in diameter, with numerous tapering, very succulent blue leaves. These leaves are up to 5 in. (12·5 cm) across at the base, slightly concave on the upper surface and very convex on the underside. Light brown horny thorn at the end of each leaf, up to 1 in. (2·5 cm) long; at regular intervals along the two leaf edges are smaller, similar coloured thorns which are curved. Flower spike up to 15 ft (5 m) or so in height, freely branched near the top, with a large number of flowers, up to 2 in. (5 cm) across, in varying shades of yellowish-green. Seed pod cylindrical, up to 2 in. (5 cm) long, dry when ripe, grey-brown.

COMMON NAME: Blue Century Plant.
PLATE: No. 97.

Agave parviflora. Torr.
ăgā'hvĭ pā'rvĭflō'ră
GROUP: *Agave.*
SYNONYM: *None.*

HABITAT: Cochise, Pima and Santa Cruz counties in Arizona and in the Chihauhua and Sonora regions of Mexico, growing on rocky hillsides usually, sometimes in grassland too.

DESCRIPTION: Low-growing stemless species, solitary, occasionally branching, and about 6 in. (15 cm) in diameter. The rosette consists of stiff narrow leaves, up to 4 in. (10 cm) long and ½ in. (1·25 cm) wide nearer the base. Leaf convex on the lower surface almost flat on the upper, dark green with white lines on top, and various white threads along the leaf edges, particularly on the upper half, whilst the end of the leaf has a very sharp, hard brown thorn. Flower spike up to 5 ft (1·5 m) high, occasionally more, branched, bearing numerous creamy-white flowers, each about ½ in. (1·25 cm) long. Seed pod about ¾ in. (2 cm) long, brown.

COMMON NAME: None.

NOT ILLUSTRATED.

Agave toumeyana. Trel.

ăgā'hvĭ tŏŏ'māyănă

GROUP: *Agave*

SYNONYMS: None.

HABITAT: Gila, Graham, Maricopa and Pinal counties in Arizona, in quite varied locations up to 5,000 ft (1,500 m).

DESCRIPTION: Stemless succulent up to 15 in. (37·5 cm) in diameter, with numerous stiff leaves. Leaves up to 10 in. (25 cm) long, less than 1 in. (2·5 cm) wide for much of their length, dark green on the upper flat surface with some white markings, green on the lower convex surface. Edges of the leaves towards the tip bear a few white threads. Leaf tip hard and sharp, brownish. Flower spike up to 8 ft (2·6 m) high, branched, bearing numerous greeny-yellow flowers just over 2 in. (5 cm) in length.

COMMON NAME: Not known.

NOT ILLUSTRATED.

(Although this species is not illustrated, it is somewhat similar to another Mexican species which appears in a view of The Exotic Collection, Worthing, Plate 138. *A. toumeyana* does not branch so freely as *A. filifera*, and the rosettes are slightly more open in appearance.

Manfreda variegata. Rose.

mănfrē'dă vă'rĭgā'htă

GROUP: *Agave*.

SYNONYM: *Agave variegata*. Jacobi.

HABITAT: Cameron, Starr, Webb and Zapata counties in Texas, also in adjoining Mexico, on the other side of the Rio Grande River, in desert-scrub areas among grass at low altitudes.

DESCRIPTION: Low growing, perennial succulent, with a tuberous root system. The rosette occasionally exceeds 2 ft (60 cm) in diameter, and consists of about 10 soft leaves which have a maximum diameter of 1½ in. (3·8 cm). These are green with brown or even purplish black irregular markings. The leaves also possess irregular teeth along the margins, but these are soft compared with those of *Agaves*. The leaves taper to a point, but do not have a hard thorn at the end as with *Agaves*. The flower spike can reach 4 ft–5 ft (1·3 m–1·6 m) in height, with 30 or so flowers, up to 2 in. (5 cm) in diameter, dark brown in colour. Seed pod, 1½ in. (3·8 cm) long and over ½ in. (1·3 cm) across, dull brown.

COMMON NAME: Not known.

NOT ILLUSTRATED.

FAMILY: BROMELIACEAE

Hechtia scariosa.

hĕ'ktĭă scărĭŏ'să

GROUP: *Bromelia*.

SYNONYMS: None.

HABITAT: Brewster county, Texas, and across the Rio Grande into nearby Mexico, on limestone hills and similar hot dry spots.

DESCRIPTION: Stemless rosettes, solitary or forming small numbers of heads, consisting of a large number of narrow, tough, recurved leaves. Rosette up to 15 in. (37·5 cm) in diameter. Leaves up to 10 in. (25 cm) long, and little more than ½ in. (1·3 cm) wide, upper surface somewhat concave. Leaf edges armed with short curved thorns, reddish-brown, at intervals of about ⅓ in. (9 mm). Leaf colour, olive-green to reddish-brown towards the upper half. Flower head is made up of numerous small flowers, close together and about 2 in. (5 cm) high, rose red.

COMMON NAME: False Agave.

Note. This plant is a relative of the Pineapple, and if you do not already know it, the formation of the leaves on top of the Pineapple will give you an idea of its appearance.

NOT ILLUSTRATED.

FAMILY: CRASSULACEAE

Dudleya brittonii. Johansen.
dŭ'dlĕă brĭ'tŏnĭē
GROUP: Echeveria.
SYNONYM: None.

HABITAT: San Diego county in California and into Baja California, Mexico, as a coastal plant on cliffs and among other vegetation on sloping ground.

DESCRIPTION: Stemless leafy succulent, usually solitary, rosettes up to 15 in. (37·5 cm) in diameter, but often only half that size. Leaves green or glaucous-white, between 3 in. and 12 in. (7·5 and 30 cm) long, 1½ in. and 3½ in. (3·8 and 8·5 cm) wide. Leaves taper to a point, up to ⅓ in. (9 mm) thick at the base. Flower spike up to 24 in. (60 cm) high, bearing a number of branches, with 10–20 flowers, just

over ½ in. (1·3 cm) in diameter, yellow.

COMMON NAME: Not known.
PLATE: No. 139.
Note: Glaucous white-leaved succulent on left of bowl.

Dudleya lanceolata. B. & R.
dŭ'dlĕă lă'nsĕŏlā'htă
GROUP: Echeveria.
SYNONYMS: *Echeveria lanceolata.* Nutt.
Cotyledon lanceolata. Brew & Wats.
Well over twelve other names also refer to this species.

HABITAT: Los Angeles, Orange, San Diego, Santa Barbara and Ventura counties in Arizona, also into the Baja Californian region of Mexico, grows as a coastal plant, on hillsides among bushes and on cliffs.

DESCRIPTION: Stemless leafy succulent, usually solitary, up to 10 in. (25 cm) in diameter, when in growth. Leaves green or glaucous, up to 8 in. (20 cm) long and just over 1 in. (2·5 cm) wide at the base, tapering to a point. Flower spike varies from 12 in.–24 in. (30–60 cm) and bears up to 20 small flowers, which are yellow or red in colour, or a mixture of the two.

Note. The coastal region of southern California, down into the Baja Californian peninsula abounds with a great variety of Dudleyas, growing on cliff faces, slopes, etc.

PLATE: No. 116.
Note: Green leaved succulent with reddish-brown tips and four flower spikes in Plate 116.

Dudleya saxosa. B. & R.
var. **collomiae.** Moran.
dŭ'dlĕă să'ksō'să cŏlŏ'mĭē
GROUP: *Echeveria.*

SYNONYMS: *Dudleya collomiae*. Rose.
Echeveria collomiae. K. & P.

HABITAT: Gila, Maricopa and Pina counties in Arizona, below 5,000 ft (1,600 m) on hillsides, among grass, bushes and on canyon ledges.

DESCRIPTION: Stemless succulent up to 6 in. (15 cm) in diameter. Leaves, which have a somewhat mealy white surface, are up to 6 in. (15 cm) in length, semi-erect, under 1 in. (2·5 cm) across at the base, tapering to a point. Flower spike about 12 in. (30 cm) long bearing up to 12 yellow flowers.

COMMON NAME: Not known.

PLATES: Nos. 100 and 140.

Note: Glaucous leaved succulent at right of bowl.

Graptopetalum rusbyi. Rose.

grǎ′ptŏp′ĕtălŭm rŭ′sbĭē

GROUP: *Echeveria*.

SYNONYMS: *Cotyledon rusbyi*. Greene.
Echeveria rusbyi. Nels & Macbr.

HABITAT: Gila, Maricopa, Pima and Pinal counties in Arizona, below 5,000 ft (1,600 m) on hillsides, among grass, bushes and on canyon ledges.

DESCRIPTION: Very dwarf, stemless succulent, rarely more than 2 in. (5 cm) in diameter. Leaves blue-green in colour, just under 1 in. (2·5 cm) in length and about ½ in. (1·3 cm) across at the base, tapering to a point. Flower spike is well branched up to 4 in. (10 cm) in height, bearing flowers just under ½ in. (1·3 cm) in diameter, off-white nearer the centre to red at the tips of the petals. Up to 24 flowers on each flower spike.

COMMON NAME: Not known.

PLATE: No. 101.

FAMILY: EUPHORBIACEAE

Euphorbia antisyphilitica. Zucc.

ūfŏr′bĭă ă′ntĭsĭfĭlĭ′tĭkă

GROUP: *Euphorbia*.

SYNONYM: *Euphorbia cerifera*. Alc.

HABITAT: Brewster, Culberson, Jeff Davis, Presidio, Terrell and Val Verde counties in Texas, and in the nearby Chihuahua region of Mexico on stony ground, in the open or beneath bushes.

DESCRIPTION: Shrubby succulent, with numerous slender grey or chalky-grey cylindrical stems, up to 4 ft (1·3 m) in height, but often only half that height, and ⅕ in. (5 mm) in diameter. The five petalled flowers, varying from white or pink to red and about ½ in. (5 mm) in diameter, appear singly or in clusters on the upper half of the stems. Seed pod or capsule is made up of three parts, almost globular, ¼ in. (6 mm) in diameter, grey when ripe.

COMMON NAME: Candelilla or Wax Plant.

PLATE: No. 48.

Note: The thin grey stems to the back of the picture, Plate 48.

Jatropha berlandieri. Torr.

jătrŏ′fă bĕrlă′ndĭĕrē

GROUP: *Euphorbia*.

SYNONYM: *Jatropha cathartica*. Teran and Berland.

HABITAT: Hidalgo, Kinney, Maverick, Starr, Val Verde, Webb and Zapata counties in Texas, and in the Coahuila and Nuevo Leon regions of Mexico, in grassland areas and desert-scrub regions.

DESCRIPTION: Tuberous rooted succulent, producing annual growth each year, with an erect stem barely ¼ in.

(6 mm) in diameter, and up to 10 in. (25 cm) high, but usually much less. This cream-coloured stem has a number of pale blue or powder-blue leaves which are intricately cut into a number of parts, held out horizontally from main stem. Flowers are orange-red in colour, up to 12, often much less, each about ⅓ in. (9 mm) in diameter. Seed pod under ½ in. (1·3 cm) almost spherical, pale brown when ripe, made up of 3 sections, containing 3 seeds, as with other members of the Euphorbiaceae family.

COMMON NAME: Not known.

NOT ILLUSTRATED.

FAMILY: FOUQUIERIACEAE

Fouquieria splendens. Engelm.

foŏkē′rĭă splĕ′ndĕns

GROUP: *Fouquieria.*

SYNONYMS: Quite a number of other names exist, as *Fouquieria* has been placed previously in other families, including Portulaceae (Purslane family) and Tamariscaceae (Tamarisk family).

HABITAT: Brewster, Crockett, Culberson, Crane, El Paso, Jeff Davis, Presidio, Reeves, Terrell and Val Verde counties in Texas, Nr. Carlsbad and the south-western counties of New Mexico, Cochise, Gila, Graham, Greenlee, Maricopa, Mohave, Pima, Pinal, Santa Cruz, Yavapai and Yuma counties in Arizona, Imperial, Riverside and San Bernardino counties in California, and over a wide area of the Chihuahua and Sonora desert regions of Mexico, in many types of habitat.

DESCRIPTION: A tree-like plant up to 25 ft (8·3 m) in height, but made up of numerous slender stems many of them of equal height, from a central basal trunk which in very old specimens can be up to 27 in. (70 cm) in diameter. Normally, average-size specimens will be one-third this size in diameter or less. The tall-growing lateral branches are usually up to 2 in. (5 cm) in diameter at their widest point, tapering towards the tip. The stems, which vary from slate-grey to greenish-brown, have ½ in.–1 in. (1·3–2·5 cm) long, slightly curved thorns at varying intervals along the entire length of the stems. Flowers in narrow clusters up to 1 in. (2·5 cm) long, orange-red.

COMMON NAME: Ocotillo or Coach Whip.

PLATES: Nos. 43, 44 and 96.

FAMILY: LILIACEAE

Dasylirion leiophyllum. Henze.

dă′sĭlē′rĭŏn lē′ŏfĭ′lŭm

GROUP: *Yucca.*

SYNONYMS: None.

HABITAT: Brewster, Jeff Davis, Presidio, Terrell and Val Verde counties in Texas, and in nearby Mexico, across the Rio Grande, in stony desert-scrub areas below 4,000 ft (1,300 m).

DESCRIPTION: Stemless or short stemmed in age, solitary or clumping, with huge rosettes up to 5 ft (1·6 m) high and 4 ft (1·3 m) in diameter. Leaves long and recurved up to 3 ft (91 cm), about 1 in. (2·5 cm) in diameter, green, with concave upper surface, and along the margins short curved thorns fairly close together. Flower spike up to 12 ft (4 m) high, unbranched, bearing 2 in. (5 cm) diameter cream flowers, which do not open wide.

COMMON NAME: Sotol, Desert Candle and Spoon Plant.

Note. The leaves when removed from the plant are spoon-shaped at the base. These leaves are often used in flower arrangements of desert plants.

PLATE: No. 17.

Note: Closely related to *D. texanum* and *D. graminifolium* from nearby regions, if in fact they are not synonymous.

Hesperaloe parviflora.

hĕ'spĕră'lō pă'rvĭflo'ră

GROUP: *Yucca.*

SYNONYMS: *Hesperaloe yuccaefolia.* Henze. *Hesperaloe engelmannii.*

HABITAT: Val Verde county, Texas.

DESCRIPTION: Stemless plant, forming into clumps consisting of narrow erect, dark green leaves, with white threads stripping from the leaf edges at intervals. Leaves up to 3 ft (91 cm) in length, occasionally more, but some specimens have only 2 ft (60 cm) long leaves, 1 in. (2·5 cm) in diameter at the widest point, and with a concave upper surface. Leaves are stiff, but slightly curved. Flower spike varies from 3 ft to 6 ft (91 cm–2 m) high, with a few horizontal branches, these bearing 1¼ in. (3 cm) long, rose-red narrow bell-shaped flowers.

COMMON NAME: Red-flowered Yucca or Samandoque.

NOT ILLUSTRATED.

Nolina erumpens. Koch.

nōlē'nă ĕrŭ'mpĕns

GROUP: *Yucca.*

SYNONYMS: None.

HABITAT: Brewster, Culberson, Hudspeth, Jeff Davis, Presidio, Terrell and Val Verde counties in Texas, particularly down near the Rio Grande, also in nearby Mexico, on grassland and hillsides below 5,000 ft (1,600 m).

DESCRIPTION: Stemless plant up to 3 ft (91 cm) in diameter, usually in clumps, with narrow almost grass-like leaves, the outer ones recurved to the ground. Leaves vary between 2 ft 6 in. and 3 ft (75 cm and 91 cm) in length, about ¾ in. (2 cm) in diameter nearer the base, whilst the margins are quite rough like Pampas grass. Flower spike partly within the rosette, about 2 ft (60 cm) long and 1 ft (30 cm) wide, with a large number of small creamy-yellow flowers forming a dense mass.

COMMON NAME: Mesa Sacahuista or Foothill Basketgrass.

NOT ILLUSTRATED.

Nolina parryi. Wats.

nōlē'nă pă'rĭē

GROUP: *Yucca.*

SYNONYM: *N. bigelovii.* P. L. Benson.

HABITAT: Orange, Riverside, San Diego and Ventura counties in California, on dry slopes below 3,000 ft (1,000 m) in Chaparral regions and even up to 5,500 ft (1,800 m) amongst Junipers and other trees.

DESCRIPTION: Grows erect usually, with a stem up to 3 ft (1 m) or more in height, bearing a dense head of green or grey-green leaves. These are quite thick, particularly towards the base, concave, with somewhat rough margins and up to 3 ft (1 m) or more in length. Flower-spike usually exceeds 4 ft (1·3 m) in height, branched, very dense in formation, and bearing a large number of creamy-white coloured flowers. These measure about $\frac{1}{2}$ in. (1·25 cm) in length, sometimes less, and about $\frac{3}{4}$ in (2 cm) wide.

COMMON NAME: Beargrass, but this name also refers to a totally different plant from further north, *Xerophyllum tenax*, outside the scope of this book.

PLATE: No. 134.

Yucca brevifolia. Engelm.

yŏŏ'kă brĕ'vĭfŏ'lĭä

GROUP: *Yucca*.

SYNONYM: *Yucca draconis* var. *arborescens* (Torr.)

Yucca arborescens (Trel.)

Clistoyucca arborescens (Trel.)

HABITAT: Inyo, Kern and San Bernardino counties in California, Mohave, Yavapai and Yuma counties in Arizona and in south western Utah, in sandy and rocky desert regions.

DESCRIPTION: Rather grotesque freely-branching tree up to 40 ft (13·3 m) in height, with a trunk up to 3 ft (91 cm) in diameter, bark brown. The upper branches are often somewhat contorted, clothed in old dead brown leaves, downward pointing. The fresh green leaves which form a dense head or rosette are up to 9 in. (23 cm) in length and about 1 in. (2·5 cm) across at the widest point. The leaf tip possesses a sharp thorn, whilst the leaf edges are rough. The flowers are over 2 in. (5 cm) across, greenish-white in colour and waxy. Each flower head, made up of a large number of flowers, can measure 14 in. (35 cm) in length and 9 in. (23 cm) or so across. Seed pod cylindrical up to 2 in. (5 cm) long, brown.

COMMON NAME: Joshua Tree.

Note. A young Joshua Tree does not begin to branch until it commences flowering. Branching follows each flower head, when one head may branch into 2, 3 or 4 new ones.

PLATES: No. 109 and 110.

Yucca thompsoniana.

yŏŏ'kă tŏ'mpsŏnĭä'hnă

GROUP: *Yucca*.

SYNONYM: None.

HABITAT: Brewster, Pecos, Terrell and Val Verde counties in Texas, and into the Chihuahua and Coahuila regions of Mexico, in stony desert regions.

DESCRIPTION: Rather similar to *Y. brevifolia*, but not growing to the same large dimensions. It is an erect branching tree, with the main barky trunk rarely exceeding 12 in. (30 cm) in diameter. The branches which are produced after a head has flowered are always held erect. They have a large number of green leaves up to 10 in. (25 cm) in length, and about 1 in. (2·5 cm) wide, tapering to a sharp point. Leaves from previous years turn brown, become down-curved, and clothe the stem for many years. Flower spike up to 5 ft (1·6 m) above the rosette or head, bearing a large number of creamy-white, waxy flowers, each about 2 in. (5 cm) in diameter. Seed pod about 2 in. (5 cm) long and $\frac{1}{2}$ in. (1·3 cm) wide, pale brown when ripe.

COMMON NAME: Thompson Yucca or Bayonet Plant.

PLATE: No. 14.

Note: Closely allied to *Y. rostrata*.

Yucca torreyi. Parent.

yŏŏ'kă tŏ'rĭĕ

GROUP: *Yucca*.

SYNONYM: *Yucca macrocarpa*.

HABITAT: Brewster, Culberson, Jeff Davis, Hudspeth, Pecos, Presidio, Reeves, Terrell and Val Verde counties, Texas, and on the other side of the Rio Grande in Mexico, over quite a wide range, in desert-scrub on stony poor soil, below 4,000 ft (1,300 m) mostly.

DESCRIPTION: Tree-like, but with few branches, often solitary, up to 15 ft (5 m) or so in height, erect, and in old specimens the trunk has been known to be as much as 20 in. (50 cm) in diameter, with a barky surface, brown. Leaves vary between 18 in. and 36 in. (45 cm and 91 cm) in length, and between 1½ in. and 2½ in. (3·8 cm and 6 cm) in diameter at their widest point. They are green, tough, with concave upper surface, and the edges of the leaves are slightly rough. Flower spike up to 3 ft (91 cm) high, somewhat conical in shape, bearing quite large flowers up to 3 in. (7·5 cm) in diameter, creamy-coloured, but the outer side of the outer petals brownish-purple.

COMMON NAME: Spanish Dagger or Torrey Yucca.

PLATE: No. 35.

Note: Intermediate hybrids also known with *Y. baccata* and *Y. treculeana.*

13. SOME PHOTOGRAPHIC HINTS

There are numerous books available on the subject of photography so that detailed information on this subject is not necessary. However, as we offer a free identification service for subscribers to the Exotic Collection monthly colour publication, we see a great many colour transparencies and photographs each year sent in from many parts of the world. Unfortunately only a small proportion of these pictures are of a sufficient quality to make identification of the plants possible. A common fault is blurring of the subject matter, not because of camera movement but usually through being too close so that the plant is badly out of focus.

There are a wide range of cameras for sale ranging in cost from a few pounds (dollars) to some in excess of £100 ($250). If you are considering buying a camera for the first time, even if you can afford the most expensive, our advice would be to buy one in the lower price bracket. A less expensive camera will usually be easier to operate, as less adjustments are necessary before you take your first picture. In fact, except for the addition of an automatic light meter, some of the more simple cameras are equivalent to the box camera of half a century ago. Cameras with automatic light meters are made in such a way that purely by altering the speed of exposure dial a coloured light or some other indicator will appear in the viewfinder, when the photograph may be taken. In other words, this type of camera has a fixed lens aperture, and the exposure speed needed will vary according to how bright the light is. By following carefully the instructions supplied with the camera, perfectly satisfactory black and white or colour pictures can be taken, within certain limitations. The usual limitation with a simple form of camera is that pictures cannot be taken closer than perhaps 5 ft or 6 ft (1·6 m or 2 m). More advanced cameras will have more controls whereby the focus can be altered, perhaps down to 3 ft (91 cm), also there will be a greater range of speeds and one or two other advantages.

Even with some of the cheaper but simple cameras, it is possible to take close-up photographs, as extra lenses (termed supplementary lenses) are obtainable. There is usually no method of seeing through the lens, to view exactly what you are taking, and so make certain that your plant is in the centre of the picture, but certain devices made of wire are supplied with the lenses to help with this. A scale is also supplied with these supplementary lenses stating which can be used, and for what distances from the object to a certain marked position on the camera. Provided instructions are carried out, it is possible to take satisfactory close-up pictures for quite a small money outlay. But, too often, we see pictures where the photographer has not been accurate enough over the distance from the plant to the camera, which has resulted in blurred—or should we say out of focus—pictures.

More expensive cameras often use the extension tube method, whereby the normal camera lens is used, but how close photographs can be taken depends on how many extension tubes or extension rings are used. Usually in these cases the object is observed on a ground glass screen, and adjustments made to ensure that the image on the ground glass is sharp or in focus. The camera is then set ready for exposing the film.

The iris of the human eye opens and closes according to the degree of light intensity to allow a certain amount or optimum amount of light into the eye on to the retina. If the opening of the iris was very small, the light would be entering less. For close-up photography, you also require the smallest aperture possible for the best results, so that the iris setting on the camera should be at $f.$ 18 or $f.$ 24, or in some cases $f.$ 32. So set it at the largest figure, depending on the camera, and this will give the smallest aperture for the iris. Again, for best results with close-up photography a light meter is a good investment, and this is supplied with full instructions. Using an average type of colour film, a setting of $f.$ 24 on the camera will mean exposures in the range of $\frac{1}{2}$–2 seconds, so this means that the camera must be on a firm base, or a tripod, as you can then alter the camera to different positions and heights yet still remain firm. When making exposures of $\frac{1}{2}$ second or longer, slight camera movement can occur on a tripod. The purchase of a cable release attachment can eliminate this problem. If you take photographs of very small flowers, the length of exposure required can be very long, sometimes 15 or 30 seconds, so camera movement and movement of the object must be avoided. To avoid movement of the object where plant life is concerned is not usually difficult, particularly if you are taking the photo in a greenhouse where there is no wind. When filming in the open-air, this can be quite a problem, and wind shields may be needed.

The accuracy of the timing mechanisms in cameras can vary, even with a brand new camera, so treat the first film or so as experimental. Also once you have graduated to the use of such extras as a light meter, you need to get familiar with it so that it becomes almost automatic. Various methods are recommended as to how to use a light meter and work out the exposure required. As a general rule it is useless to take your light reading off a white or near-white object; it is far better to do it from a green leaf or something similar. Never aim the light meter intentionally or inadvertently at the sun, as it is the best way to damage the light cells and cause inaccuracies. If you are out in desert regions, the light will be very bright indeed, and it can be difficult to get accurate light readings. As a general rule, but remember this is only a theoretical example as it will depend on shutter speed, and film speed, if your light meter suggested an aperture setting of f. 14, you will usually find that f. 15 to f. 16 will give best results. In hot desert type regions it is very easy to over-expose film, which will result in colour transparencies with the colours being a little paler than they should be.

When filming very small plants and flowers, requiring the use of additional extension tubes or rings, or supplementary lenses, it is necessary to multiply the exposure time by two, three or four times, according to how close the camera is to the object. Although guidance is given on these ratio factors when purchasing these extras for close-up photography, an experimental film or so using varying exposures at varying distances should be taken. When doing this, record the light factor and the length of exposure given each time, and then once the film has been processed you can then make your own judgments of any exposure differences necessary for the future from the results of such a film.

When taking close-up photographs of plants, try to select a good background, so that not only does the flower show up well but also the plant details. A black background can be very impressive, but if your cactus plant has long black spines, they will tend to merge in with the background of black. With an example of this sort, it is sometimes still possible to use the black background with the careful use of pale coloured rocks behind the plant, and the flower must stand up sufficiently above them. Blue sky can make a wonderful background too. The colour backing does not really matter, provided the important features of the plant and flower are still clearly visible. Some people favour a blurred background which you get if you film a cactus among dried grasses in its natural habitat. In this case you will have a very sharp and clearly defined cactus against a backing of out-of-focus grasses.

If you can obtain a three-dimensional finished result, no matter whether it is a black and white or a colour transparency for projection, it will give a much better idea of the subject matter. The angle of light can help a great deal, so try to get the light which is illuminating the subject to come slightly from an angle, rather than straight at the object. Slight shadows help to give a three-dimensional effect, but they do not want to be so dark as to obscure detail, unless it cannot be avoided.

For those taking colour photographs, we would recommend the permanent use of an ultra-violet or haze filter. It is usually suggested for special use when filming in snow conditions, over the sea, etc., to try to avoid anything white having a tinge of blue in it. We have found that even when taking close-up pictures of small flowers with white hairs we have obtained a much purer white on those hairs with this filter over the lens. It also has another advantage; it protects the actual lens of the camera from becoming soiled or dusty. It is easy to clean a glass filter, whereas a camera lens must be cleaned very carefully. If the glass filter eventually becomes scratched, it is quite an inexpensive item to be replaced, but to replace the camera lens will cost a lot of money.

In order to obtain a fairly accurate colour record, another useful tip concerns taking photographs of flowers in the magenta to purple range. If these are taken in full sun, the flowers will tend to come out more red than they should do. To avoid this, flowers in these colours should be filmed either in natural shade, when clouds obscure any direct sunlight, or by placing an opaque shield between the sun's rays and the flower to produce the same effect.

We have not mentioned types of cameras, as this is a matter of personal choice. There are varying film sizes, and by far the cheapest to operate is a camera using 35 mm film. You can go smaller still, but for good quality pictures, particularly if you want transparencies for projection purposes, 35 mm size takes a lot of beating, and there is also a wide choice of cameras and films.

Just to finish, here is a short list of the basic needs for taking close-up pictures, no matter whether this is to be done in the open-air or in the confines of a greenhouse. We have avoided mentioning the use of flash-light methods, as the majority of species, including the night flowering plants, can usually be photographed in daylight in the early morning and still give very good results.

REQUIREMENTS FOR CLOSE-UP PHOTOGRAPHY

1 medium priced 35 mm camera.

1 tripod.

1 light meter.

1 cable-release.

1 ultra-violet or haze filter.

plus extension tubes or supplementary lenses according to the camera you buy, in order to get down 12 in.–15 in. (30 cm–37·5 cm) or less from the object.

14. THE EXOTIC COLLECTION

This private botanical garden contains over 9,000 species of cacti and other succulent plants, with five specially laid out greenhouses ranging from our unheated experimental house to other heated ones, including one for the very tropical species. In addition there are rockeries outside for the cacti and other succulent plants which can be grown in the open air throughout the year. EVERY MONTH, The Exotic Collection sends its subscribers TWO NEW (previously unpublished) PHOTOGRAPHIC REFERENCE PLATES in COLOUR. (Size $8\frac{1}{2} \times 6$ in.) (23 cm $\times$ 15 cm), with non-technical cultural notes, etc.

AN EIGHT PAGE Monthly Notes (also illustrated in COLOUR). (A total for one year of 24 plates and 96 pages of Monthly Notes—minimum 72 pages of FULL COLOUR, together with articles and other cultural information. Some of these colour illustrations will cover two pages filmed by us here in the collection or in habitat. OVERSEAS subscribers also receive an additional 4-page Overseas Newsletter, containing specialised cultural information for varying countries, and this is usually issued bi-monthly.

ALL SUBSCRIBERS receive a seeds list of named species, some of which are free. Plants also available to subscribers.

SUBSCRIPTION operates from January to December each year. If you join during any year you will automatically receive all the back issues for the current year to the month of joining. After that future issues will be posted during the first week of each month.

SUBSCRIPTION for ONE YEAR — £2·30 for the UK and Ireland.

SUBSCRIPTION for ONE YEAR — $6·00 (USA) or equivalent for all overseas countries.

Under the personal direction of
Edgar Lamb and Brian M. Lamb.

Letters to
16, Franklin Rd,
WORTHING, Sussex,
BN13 2PQ, England.

15. SIMPLIFIED GLOSSARY OF BOTANICAL TERMS

(excluding any already explained)

AREOLE   In cacti, it is the position or area from which the spines (and with *Opuntias* the glochids as well) develop. Flowers also appear either from the areole or from just above it as with the *Echinocerei.*

DECIDUOUS   Deciduous leaves are those which fall each season, rather than remaining on the tree for a number of seasons, as with evergreens.

EPIDERMIS   The outer layer of cells on the body or leaves of plants, which with many desert plants have waxy surfaces, which helps reduce the water loss.

FAMILY   A group of closely related tribes or genera, belonging under one heading, and this word usually ends in *aceae.*

FORMA   Often abbreviated to *fa.* A lower or less important division of plants, below the rank of variety or varietal difference.

GENUS (plural = genera.)   A subdivision of a Family, consisting of one or more species. These species have certain common factors or characteristics. The genus is the first name of a plant, commencing with a capital letter.

GLAUCOUS   Bluish-grey or bluish-green; covered with a whitish powdery bloom as on the surface of a grape.

GLOCHID   Small barbed bristle or hair, usually in clusters, and appearing from an areole, and occurring in the genus *Opuntia.*

INFERIOR OVARY   The ovary is the organ or part of a flower in which the seeds can form, and when it is inferior, it is below the other parts of the flower, such as the petals, stamens, etc.

PECTINATE   Referred to here in reference to spine formation, meaning comb-like, like the row of teeth on a comb.

PETAL   The parts of the flower, which are usually highly coloured.

RADIAL   *See* Spine.

RIB   Used in this book with reference to the formation of the

FLOWER STRUCTURE

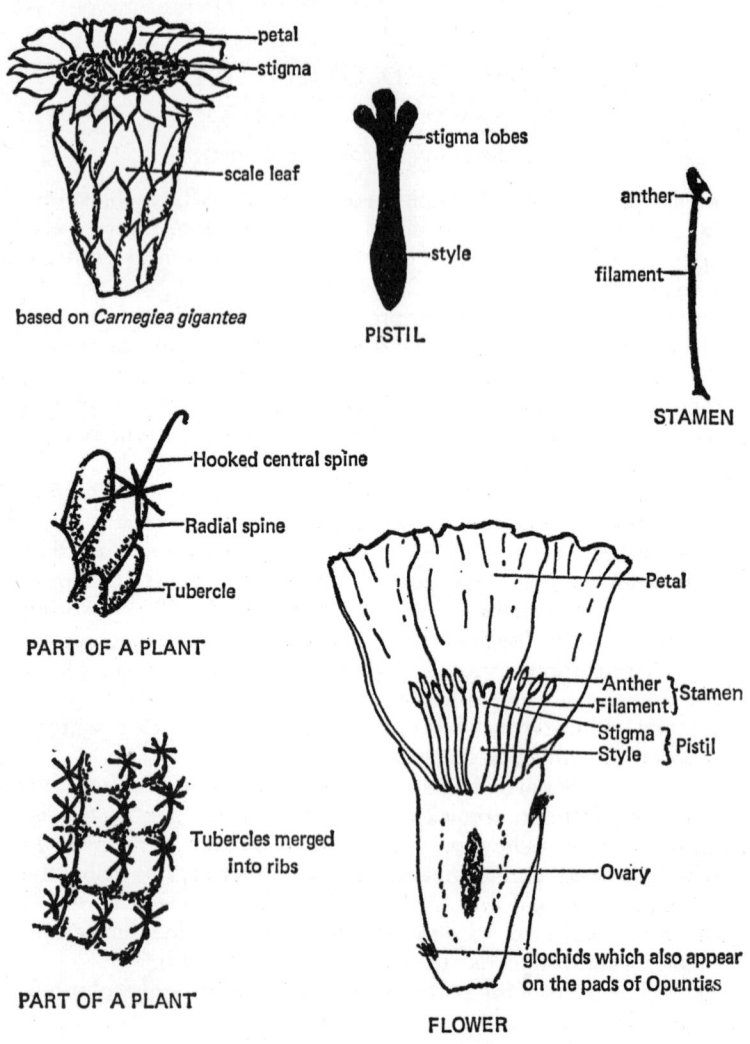

petal
stigma
scale leaf

based on *Carnegiea gigantea*

stigma lobes
style

PISTIL

anther
filament

STAMEN

Hooked central spine
Radial spine
Tubercle

PART OF A PLANT

Tubercles merged
into ribs

PART OF A PLANT

Petal

Anther } Stamen
Filament }
Stigma } Pistil
Style }

Ovary

glochids which also appear
on the pads of Opuntias

FLOWER

based on *Opuntia engelmannii*

SPINE FORMATIONS

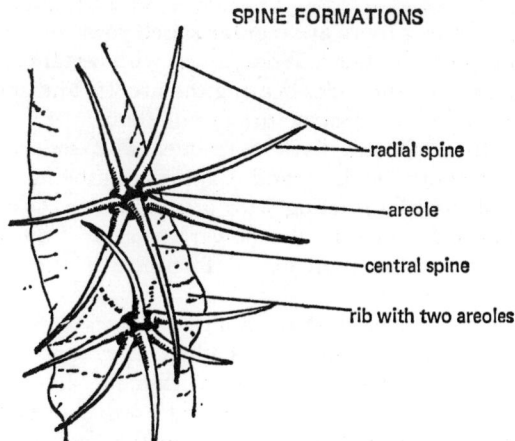

radial spine

areole

central spine

rib with two areoles

based on *Echinocactus horizonthalonius*

pectinate spines

central spine

radial spine

two tubercles bearing
two areoles from
which spines appear

based on *Neolloydia conoidea*

TUBEROUS TYPE ROOT SYSTEMS

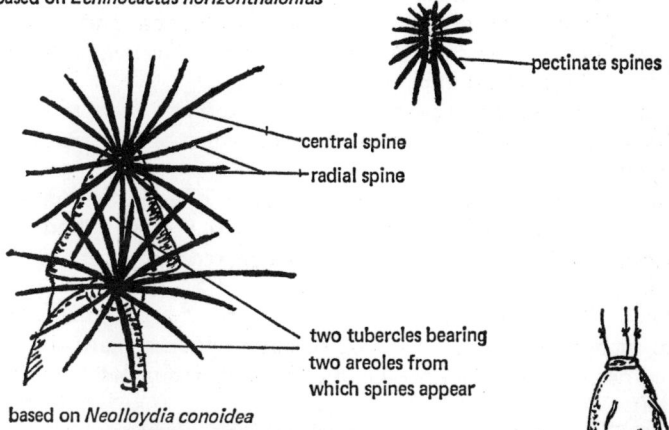

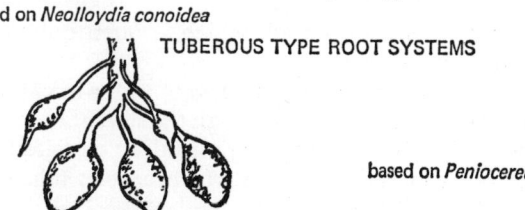

based on *Peniocereus*

based on *Wilcoxia* and *Neoevansia*

stems of cacti, whereby there are regular raised rows, on which the areoles and spines are borne. Sometimes, with certain cacti, as they grow older, the tubercles bearing the areoles and spines merge together to form a rib-structure on the stems.

SPECIES   This is the first sub-division or means of separating plants under one genus heading, and it appears in the name as the second word, usually starting with a small letter. Usually this means some difference in the flower structure from one species to another, rather than just a difference in the plant structure.

SPINES   In cacti they are usually sharp-pointed objects appearing from an areole, some times singly, but often in groups. They are formed from leaves made by a bud within the areole. In some cases the spines can be separated into *central* ones and *radials*. The central ones are obviously from the centre of an areole, while the radials are those from the edge of the areole.

STAMENS   These are the male parts of a flower, made up of two parts. The basal part or stalk is known as the *filament*, on top of which is the *anther*, that bears the pollen.

STERILE   This means imperfect, and in this book refers to fruits which either contain no seeds, or seeds which are incapable of germinating.

STIGMA   Among the stamens in a flower, and often in the middle, and sometimes raised above the stamens you will find the *stigma*, which is the top part and pollen receiving part of the *pistil*.

STOMATA   This is the plural form of the word *stoma*, a pore in the leaf or body surface of a plant, through which a plant simply breathes, and by which water can be transpired or lost from the plant.

TESTA   The outer surface of a seed, which is often quite hard, particularly with cacti seeds.

TRANSPIRATION   The term used for the loss of water in vaporised form, from a plant, which occurs through the stomata.

TUBERCLE   In cacti, tubercles are the projections or raised positions on the body of the plant, on which areoles and spines appear.

TUBEROUS   A modified form of root, whereby it is swollen and has become a storage organ.

VARIETY   Usually written in the abbreviated form *var.* is often the third word in a plant name, and is a further means of separating one plant from another, but of far less importance than a species difference.

Note   We have deliberately minimised the possible number of terms used in this book, to simplify it for readers. For example, there are various types of spines, in structure or angle of projection from the plant, and these often have special terms, but we have avoided using them.

ACKNOWLEDGMENTS

We should like to thank the following friends, who gave us great assistance in the field work we carried out in Texas, Arizona and California, and also for their help at other times over a period of many years:

Don and Eloise Johnson (Corpus Christi)
Alan and Betty Blackburn (Tucson)
Paul Shaw (Tucson)
Gilbert Voss (Encinitas)
Dr Philip Corliss (San Diego)
Don and Chauncey Cox (Corpus Christi)
Mike and Anne Soos (Clifton)
Homer Jones (Alpine), a wonderful 'old-timer' who knew more about the cacti of the Big Bend area of Texas than anyone else with whom we have corresponded and with whom we spent many enjoyable hours. Very sadly he died just over a year before the publication of this book.

We received similar help from the Arizona-Sonora Desert Museum (Tucson). We should like to thank Alan Blackburn for the loan of a number of transparencies. Special thanks must go to Dr George Lindsay, Director of the California Academy of Sciences in San Francisco, for the loan of two transparencies and his advice and help in checking for accuracy. We should also like to thank the United States of America National Parks Service and the respective Superintendents for their assistance in checking the accuracy of the chapter on National Parks and Monuments, and for the loan of three transparencies, one filmed within the Saguaro National Monument, the other two within the Joshua Tree National Monument. Finally we must thank for their assistance those ranchers in Texas who, for conservation reasons, wish to remain anonymous.

The following books were very useful in checking the common names and certain habitats when the manuscript was being prepared.

Cacti of the Southwest Hubert Earle
The Cacti of Arizona Lyman Benson
The Native Cacti of California Lyman Benson
Cacti of the Southwest Del Weniger
Wildflowers of the Big Bend Country, Texas B. H. Warnock
The Cactaceae Britton and Rose
Saguaro National Monument Napier Shelton (National Park Service)
Wildflowers of the USA Rickett
A California Flora P. A. Munz

INDEX OF LATIN NAMES

Numbers in roman type refer to pages; numbers in bold type are illustration numbers.

INDEX OF ENGLISH NAMES